WHELAN LAKE (CA-SDI-6010): A LA JOLLAN CAMPSITE ON THE LOWER SAN LUIS REY RIVER, SAN DIEGO COUNTY, CALIFORNIA

by

Rein Vanderpot
Jeffrey H. Altschul
Donn R. Grenda

with contributions by:

Paul D. Bouey
Kellie M. Cairns
Owen K. Davis
C. Kimberly Green McClure
Charles H. Miksicek
Thomas Origer
Anthony Orme
Keith J. Vlastos

Statistical Research, Inc.
Technical Series, No. 40
1993

ISBN 1-879442-38-8

Statistical Research, Inc.
P.O. Box 31865
Tucson, AZ 85751

TABLE OF CONTENTS

LIST OF FIGURES ix

LIST OF TABLES xi

ACKNOWLEDGEMENTS xiii

ABSTRACT xv

CHAPTER 1: INTRODUCTION 1
REPORT ORGANIZATION 5

CHAPTER 2: ENVIRONMENTAL BACKGROUND 7
INTRODUCTION 7
GEOLOGY 8
SOILS 8
CLIMATE 9
HYDROLOGY 10
VEGETATION 10
FAUNA 13
Avifauna 14
Fauna In Prehistory 16
CONCLUSIONS 17
A GEOMORPHOLOGICAL ASSESMENT OF THE STUDY AREA
by Antony R. Orme 18
Regional Setting: The San Luis Rey River Basin 18
The Drainage Basin 18
Geologic Framework 19
Hydrology and Flood History 19
The Whelan Lake Area 21

CHAPTER 3: CURRENT ISSUES IN CULTURE HISTORY 25
INTRODUCTION 25
PREVIOUS ARCHAEOLOGICAL INVESTIGATIONS 25
Early Archaeological Investigations 1920 - 1955 27
The Middle Years 1955 - 1972 27
Recent Archaeological Investigations 1973 - Present 28
ESTABLISHING CHRONOLOGY 31
Terminology 31
The Sequence Begins 33
Early Man? 35
Hunting Cultures 35
The Transition Period from Hunting to Food Collecting 36
Food Collecting Cultures (La Jolla, Pauma, and Sayles) 36
Diversified Subsistence (Late Prehistoric Cultures) 37
THE "CULTURE" CONCEPT AS USED
IN CALIFORNIA ARCHAEOLOGY 37
Hunting Cultures 38
Food Collecting Cultures 39
La Jolla Culture 39

Pauma Culture........ 40
Sayles Culture........ 40
Diversified Subsistence (Late Prehistoric Cultures)........ 41
The San Luis Rey Culture........ 41
The Irvine Culture........ 42
The Cuyamaca Culture........ 42
Integrating Settlement, Subsistence, and Chronological Issues in Prehistory........ 42
HISTORY OF THE SAN LUIS REY REGION........ 43
Ethnohistoric and Ethnographic Background........ 43
Mission San Luis Rey........ 46
The Rancho Period........ 46

CHAPTER 4: RESEARCH DESIGN........ 47
INTRODUCTION........ 47
THE SCIENTIFIC METHOD IN ARCHAEOLOGY........ 47
RESEARCH QUESTIONS AND DISCUSSIONS........ 48
General Questions........ 48
General Methodology........ 50
Specific Prehistoric Research Domains........ 50
Paleoenvironmental Issues........ 50
Research Questions........ 50
Data Requirements and Methodology........ 51
Chronology........ 51
Research Questions........ 51
Data Requirements and Methodology........ 52
Subsistence and Settlement........ 53
Research Questions........ 53
Data Requirements and Methodology........ 53
Site Structure and Social Organization........ 55
Research Questions........ 55
Data Requirements and Methodology........ 56
Exchange and Intergroup Relations........ 56
Research Questions........ 57
Data Requirements and Methodology........ 57
CONCLUDING REMARKS........ 57

CHAPTER 5: FIELD RESULTS........ 59
PROJECT AREA AND TEST LOCI........ 59
EXCAVATIONS........ 61
Backhoe Trenches........ 61
Locus A........ 62
Site Stratigraphy........ 65
Disturbances........ 68
Locus B........ 69
Locus C........ 71
Test Pits........ 71
Goals and Methods........ 71
Test Pit Evaluation........ 74
Clayey Areas in the Midden Fill........ 76
Feature Descriptions........ 77
Feature 1........ 77
Excavation Strategy........ 77

Description........ 78
Stratigraphy........ 78
Feature Fill and Artifacts........ 80
Interpretation........ 80
Feature 2........ 80
Excavation Strategy........ 80
Description........ 81
Feature Fill and Artifacts........ 82
Interpretation........ 82
Feature 3........ 83
Feature 4........ 84
Summary and Conclusions........ 84

CHAPTER 6: CHRONOMETRIC ANALYSIS........ 87
RADIOCARBON DETERMINATIONS FROM SHELL SAMPLES........ 87
OBSIDIAN HYDRATION ANALYSIS
Paul D. Bouey, Thomas Origer and Rein Vanderpot........ 88
Conclusions........ 91

CHAPTER 7: LITHIC ANALYSIS
Keith J. Vlastos........ 93
FLAKED STONE........ 93
Methods........ 93
Results........ 94
Raw Materials........ 94
Cortex........ 96
Flake Size........ 96
Technologies........ 98
Cores........ 99
Tools........ 100
Edge Modified Flakes........ 100
Hammerstones........ 103
Choppers........ 103
Biface Tools........ 103
Flaked Stone Distribution........ 104
Flaked Stone Conclusions........ 111
GROUND STONE........ 112
Ground Stone Conclusions........ 114

CHAPTER 8: INVERTEBRATE FAUNAL ANALYSIS
C. Kimberly Greene McClure and Jeffrey H. Altschul........ 115
METHODS........ 115
SHELLFISH IDENTIFICATION AND HABITAT RECONSTRUCTION........ 116
INTRASITE ANALYSIS........ 119
INTERSITE ANALYSIS........ 123
CONCLUSIONS........ 129

CHAPTER 9: FAUNAL ANALYSIS
Kellie M. Cairns and Jeffrey H. Altschul........ 131
INTRODUCTION........ 131
METHODS........ 131
THE ARCHAEOLOGICAL FAUNA........ 132
Mammals........ 132

Reptiles.. 137
Fish.. 137
Birds.. 138
Unidentifiable Bone.. 140
EXPLOITED HABITATS AND HUNTING PRACTICES.. 140
Animal Processing.. 141
Intrasite Analysis.. 143
INTRASITE ANALYSIS AND LA JOLLAN ADAPTATION.. 143

CHAPTER 10: MICRO- AND MACROBOTANICAL ANALYSIS.. 151
POLLEN ANALYSIS Owen K. Davis.. 151
Introduction.. 151
Methods.. 152
Results.. 152
Comparison of Archaeological Samples with the Modern Analog.. 152
Items # 96 and # 272.. 156
Items # 133 and # 125.. 156
Item # 295.. 156
Interpretation.. 157
Environment at the Time of Occupation.. 157
Ethnobotanical Information.. 157
Conclusions.. 157
MACROBOTANICAL ANALYSIS Charles H. Miksicek.. 158
Introduction.. 158
Methods.. 158
Results and Discussion.. 159
Summary.. 165
CONCLUSIONS.. 166

CHAPTER 11: ANCIENT SETTLEMENT AND SUBSISTENCE ALONG THE SAN LUIS REY RIVER.. 167
THE WHELAN LAKE SITE (CA-SDI-6010).. 167
Temporal Context.. 167
Site Layout.. 168
Feature Types.. 168
Clayey Areas of Unknown Cultural Origin.. 168
Distributional Patterns.. 169
General Site Constituents.. 169
Conclusion.. 170
REGIONAL TRENDS IN THE SAN LUIS REY RIVER BASIN.. 171
Physiography and Geography.. 171
The Lower Basin.. 171
Coastal Area.. 171
The Estuarine Zone.. 173
Freshwater Marshland.. 175
Alluvial Valley.. 175
The Middle Basin.. 176
The Upper Basin.. 177
Summary.. 177
Settlement Patterns and Intrabasin Dynamics.. 177
Lower Basin Settlement.. 178
Site Location.. 178
Site Composition.. 182

Site Function........ 183
Cultural Affiliation........ 185
Discussion........ 187
Middle Basin Settlement........ 189
Lower and Middle Basin Comparison........ 191
UNRESOLVED ISSUES........ 194

CHAPTER 12: SUMMARY AND RECOMMENDATIONS........ 195
NATIONAL REGISTER ELIGIBILITY........ 195
HUMAN REMAINS........ 197
TREATMENT PLAN........ 197
DIRECTIONS FOR FUTURE RESEARCH........ 197

REFERENCES........ 199

LIST OF FIGURES

Figure 1. The southern California coastal region 2
Figure 2. The San Luis Rey River Basin 3
Figure 3. Map of the project area, showing the Whelan Lake site and test loci 4
Figure 4. Local geologic formations 9
Figure 5. The lower San Luis Rey River basin 22
Figure 6. Regional map showing site locations 26
Figure 7. Map of California culture areas 32
Figure 8. Map of Luiseno territory and surrounding area 45
Figure 9. Photograph of project area looking south 60
Figure 10. Photograph of project area looking northwest from Locus A 60
Figure 11. Map of CA-SDI-6010, the Whelan Lake site, showing Loci A and B 64
Figure 12. Digram showing depth of shell midden across SDI-6010 67
Figure 13. East profile of Trench M 67
Figure 14. North profile of Trench H, showing irrigation pipe disturbance 68
Figure 15. One meter section of west profile of Trench K 70
Figure 16. One meter section of west profile of Trench P 70
Figure 17. One meter section of south profile of Trench V 72
Figure 18. One meter section of south profile of Trench W 72
Figure 19. South Profile of Trench G showing Feature 1 77
Figure 20. Photograph of Feature 1 as seen in Test Pit 2, Level 3 79
Figure 21. Plan map of Test Pit 2, showing Feature 1 79
Figure 22. South profile of Trench G showing Feature 2 81
Figure 23. Plan map of Test Pit 12, base of Level 2, Feature 2 82
Figure 24. Feature 3, displayed in the north profile of Trench E 83
Figure 25. North profile of Trench C 85
Figure 26. End Scraper (a) and Side Scraper (b) 102
Figure 27. Pic from Trench A 102
Figure 28. Biface tool fragments 104
Figure 29. Lithic density 107
Figure 30. Ratio of basalt and quartzite to other materials 108
Figure 31. Ratio of shatter to flakes 109
Figure 32. Ratio of other types of flakes to biface flakes 110
Figure 33. Shaped Manos 113
Figure 34. Worked bones from SDI-6010 143
Figure 35. Map of southern California showing the Whalen Lake site 154
Figure 36. Percentage Pollen Diagram 156
Figure 37. Map of the San Luis Rey River basin 172
Figure 38. Sea Level changes after 13,00 CYBP 174
Figure 39. Map of the lower San Luis Rey River basin 179
Figure 40. Lower basin sites identified by cultural affiliation 184
Figure 41. The middle basin of the San Luis Rey River 190

LIST OF TABLES

Table 1. List of Wetland Vegetation Types Near Whelan Lake ... 11
Table 2. List of Upland Vegetation Types Near Whelan Lake ... 12
Table 3. Paleoethnobotany of the San Luis Rey Area ... 13
Table 4. Observed Avifauna on the Whelan Property ... 14
Table 5. Identified Remains from SDI-5130 and SDI-6014 ... 16
Table 6. Commonly Used Cultural Sequences in the San Diego Area ... 33
Table 7. Summary of Data for All Backhoe Trenches in the Project Area ... 63
Table 8. Surface Area, Depth, and Volume of Test Pits ... 73
Table 9. Artifact Counts and Ecofact NISP per Level for Each Test Pit ... 75
Table 10. Radiocarbon Determinations for Shell Samples ... 88
Table 11. Trace Element Analysis of Obsidian Artifacts ... 89
Table 12. Summary of Hydration Band Analysis of Obsidian ... 90
Table 13. General Distribution of Raw Materials ... 94
Table 14. Amount of Cortex Present in Debitage ... 96
Table 15. Proportion of Flakes in Debitage ... 97
Table 16. Flake Size ... 98
Table 17. Distribution of Diagnostic Flakes in Each Raw Material ... 99
Table 18. Tool Distribution ... 101
Table 19. Vertical Distribution of Raw Materials ... 105
Table 20. Vertical Proportions of Raw Materials ... 106
Table 21. Vertical Distribution of Diagnostic Flakes ... 106
Table 22. Ground Stone Distribution ... 112
Table 23. Total Shell NISP by Test Pit ... 116
Table 24. Shellfish Species Identified at SDI-6010 ... 117
Table 25. Total Shell Weight (in g) by Test Pit and Feature ... 121
Table 26. MNI by Test Pit and Feature ... 122
Table 27. Chi Square Matrix and Score ... 123
Table 28. Total Shell Weight (in g) by Level ... 124
Table 29. Percentage of Shell Weight by Level ... 125
Table 30. MNI by Level ... 126
Table 31. Percentage of MNI by Level ... 127
Table 32. Intrasite Shell Comparison, Percentage of Total for Each Site ... 129
Table 33. Site SDI-6010 Species List ... 133
Table 34. SDI-6010 Frequency and Percentage of all Vertebrates ... 134
Table 35. Burned Bones Recovered from CA-SDI-6010 ... 142
Table 36. NISP by Provenience ... 144
Table 37. Comparison of MNI in the Faunal Collections ... 148
Table 38. Analyzed Pollen Samples from the Whelan Lake Site ... 152
Table 39. Pollen Extraction Procedure ... 153
Table 40. Pollen Percentages for CA-SDI-6010 ... 155
Table 41. Plant Taxa Identified from CA-SDI-6010 ... 160
Table 42. Plant Remains Recovered by Flotation from CA-SDI-6010 ... 161
Table 43. Charred Plant Remains from Sites Along the Southern California Coast ... 164
Table 44. Artifact Counts and Ecofact NISP per Level for Each Test Pit ... 169
Table 45. NISP Percentages of Ecofacts at the Whelan Lake Site ... 170
Table 46. Lithic Tool Ratios for the Whelan Lake Site ... 170
Table 47. Prehistoric Sites Recorded in the Lower San Luis Rey River Basin ... 180
Table 48. Food Sources for the Historic Luiseno ... 193
Table 49. Comparison of Artifact Frequencies ... 193

ACKNOWLEDGMENTS

A number of people deserve recognition for their contribution to this report. First and foremost, we would like to acknowledge the project sponsor, the Los Angeles District, U.S. Army Corps of Engineers. We thank in particular Stephen Dibble for his support and continued interest. In his role of administering the contract, he reviewed the draft report and supplied helpful comments. He also furnished useful maps and aerial photographs of the project area.

No one probably knows the archaeology of the San Luis Rey River basin as well as Delbert L. True and we thank him for taking the time to review the draft report and examine some of the lithic artifacts from the Whelan Lake site. This report has benefited from his thoughtful observations and suggestions. Other archaeologists who have contributed to our study include Michael Moratto, Dennis Gallegos, and John Cook.

Henry Rodriguez, representing local Native Americans, visited the site several times during the course of the project. To him we express our sincere gratitude. T. Knox Bell provided historic data on the Whelan Lake Ranch and Bird Sanctuary, thus furthering our understanding of the local environment.

The success of the project is largely due to the field crew and laboratory technicians. Donn Grenda served as crew chief, helped by crew members Patrick Clark, Matt Finney, C. Kimberly Greene McClure, Andrea Kayser, Doug McIntosh, Debbie McLean, and Keith J. Vlastos. Processing of artifacts was carried out by Kellie M. Cairns and Keith J. Vlastos under the direction of Carol Ellick. Stanley Olsen from the University of Arizona was so kind to let us use the comparative collection of faunal materials located in the zooarchaeology laboratory of the Arizona State Museum. Roy A. Salls of the Lompoc Museum identified an additional number of piscine and mammalian species not present in that collection.

Finally, we are indebted to the SRI support staff for an excellent job in producing the final report. Carol Ellick designed the report cover and layout. Illustrations and maps were drawn by Susan Martin. Michael S. Pellerin formatted and produced the final report. Sylvia Lindsay proofread the final report and served as technical editor. To all those associated with this project, we offer our sincere appreciation.

ABSTRACT

This report details the results of archaeological test excavations conducted in August, 1991, by Statistical Research, Inc. at the Whelan Lake site (CA-SDI-6010). Whelan Lake is an early La Jollan campsite, dating between 6500 and 7400 B.P. The site is a moderately dense shell midden situated on a knoll overlooking the San Luis Rey River about 7 km upstream from the coast. The midden has a roughly oval shape and measures 90 m by 60 m. Over 200 m of backhoe trenches and 17 m^2 of test pits were excavated in this area. Four cultural features, all interpreted as hearths, were exposed in the excavation units.

An accurate reconstruction of the paleoenvironment of the lower San Luis Rey River basin is crucial for a correct interpretation of the recovered data. Although the precise maximum extent of the estuarine zone during the Early Holocene is unknown. Geomorphic assessment suggests that a relatively stable, brackish lagoon once extended 2 km inland, confined to the fjord-like lower stretch of the river. A more substantial saltwater marsh may have briefly existed as far as 4 km inland. Freshwater marshes formed upstream as waters backed up behind the estuary. In agreement with the geomorphic evidence, micro- and macrobotanical analyses suggest that the site was situated in a freshwater marshland environment.

Although shellfish comprises the bulk of faunal remains present in the midden, the dietary significance of this food source is minimal. NISP for small terrestrial animals is relatively large, highlighting the importance of non-aquatic sources for meat in the diet of the site's occupants. The archaeobotanical assemblage is dominated by a mixture of edible grasses, legumes, and small seeds similar to that found at many other sites in southern California. The lithic collection reflects a hunting-gathering economy. Results from our excavations at Whelan Lake indicate that La Jollan reliance on estuarine sustenance is not always as strong as assumed. At this site, the dominant economic focus appears to have been on the freshwater marshland just above the estuary.

Reconciling the excavation data with the geomorphic evidence leads to a model of early coastal adaptation for the river that is then placed in a regional context. A comparison of recorded sites in the lower basin sheds light on the puzzling relationship between La Jollan and Pauman groups. Sites either show a mixed marine/terrestrial economy (La Jolla) or a pure terrestrial economy (Pauma). Perhaps the difference reflects cyclical exploitation of multiple ecozones by one people, rather than a disparate economic emphasis by two groups. Concentrations of groups around the estuarine zone and in the mountainous inland zone may be seen as two poles of one adaptation system. The adaptational balance shifted through time due to environmental constraints. It seems likely that when the coastal lagoon silted in, some 3,500 years ago, people gradually moved inland to exploit a more varied resource zone.

CHAPTER 1
INTRODUCTION

The San Luis Rey River basin is one of a number of east-west trending river valleys extending from the coastal range to the coast between Long Beach and San Diego (Figure 1). In general, these river basins are relatively small. The San Luis Rey drainage encompasses 1,400 km^2, with its headwaters in the Bucksnort Mountains and flowing east to the coast, a distance of 100 km (Figure 2). These small physiographic units host a diverse set of environmental resources, ranging from oak forests to ocean beaches, in a circumscribed area. People have found these basins attractive for thousands of years, presumably moving in time with the resources from one end of the basin to the other.

Archaeologists have also found these basins attractive subjects of inquiry. Viewed as laboratories, these basins provide a context from which human adaptation to coastal and inland resources can be studied. The interplay between these diverse resources has been the source of much debate. Some archaeologists interpret sites in different parts of the basin as members of different cultures, whereas others view them as different aspects of an integrated adaptation to the whole.

The San Luis Rey River basin has been a major focus of this debate. Systematic archaeological work has been conducted in the basin since the 1950s. Interpretations concerning chronology and culture that have affected all archaeologists working along the southern coast of California have derived from work in this river basin. The prehistory of the San Luis Rey River basin is a story that continues to unfold. It is our story, too.

The Whelan Lake site (CA-SDI-6010, hereafter the prefix CA is dropped) lies on a prominent knoll overlooking the north bank of the San Luis Rey River (Figure 3). The site is located approximately 7 km from the current mouth of the river, where it meets the Pacific Ocean at present-day Oceanside. The site was recorded in 1978, and has subsequently been subject to a number of limited testing projects. The results of these projects indicated that the site contained two components, a relatively recent historic occupation and a prehistoric occupation of undetermined age.

In 1991, the Los Angeles District of the U.S. Army Corps of Engineers (COE) contracted with Statistical Research, Inc. (SRI) to conduct archaeological test excavations at the Whelan Lake site. The work was conducted as part of the COE's continuing program to provide flood control to the lower reaches of the San Luis Rey River. The parcel of land owned by the City of Oceanside that includes the Whelan Lake site was considered as a potential borrow pit. To comply with various Federal mandates, including the National Historic Preservation Act and the Archaeological and Historic Preservation Act, amending the Reservoir Salvage Act of 1960, the COE had to determine whether one or both components at SDI-6010 was eligible for inclusion in the National Register of Historic Places. Based on this determination, a treatment plan for the site was developed (Altschul 1991).

The project was divided into two parts. Greenwood and Associates was retained to conduct the testing of the historic component, and their results are presented in a separate report (Schmidt et al. 1992). SRI undertook the prehistoric evaluation. At first, this part of the project seemed relatively simple. The prehistoric component is confined to a small, discrete area on top of the knoll. The site is located within 1.5 km of the San Luis Rey Mission. This fact, combined with the discovery of two brownware sherds on the surface of the site, indicated a late occupation, perhaps associated with the mission. During fieldwork, our opinion changed. Artifacts examined in the field suggested an earlier date of occupation. How early we could not tell. Indeed, it was not until the radiocarbon dates were returned that we learned the site was not one of the youngest aboriginal sites in the basin, but one of the oldest. Turning our intuitive impression on its head has led us down a number of analytical paths, some with dead-ends, others more fruitfully followed. We hope by presenting all our efforts in a straightforward manner, we can help others avoid our pitfalls, and add an important piece to the puzzle of San Luis Rey.

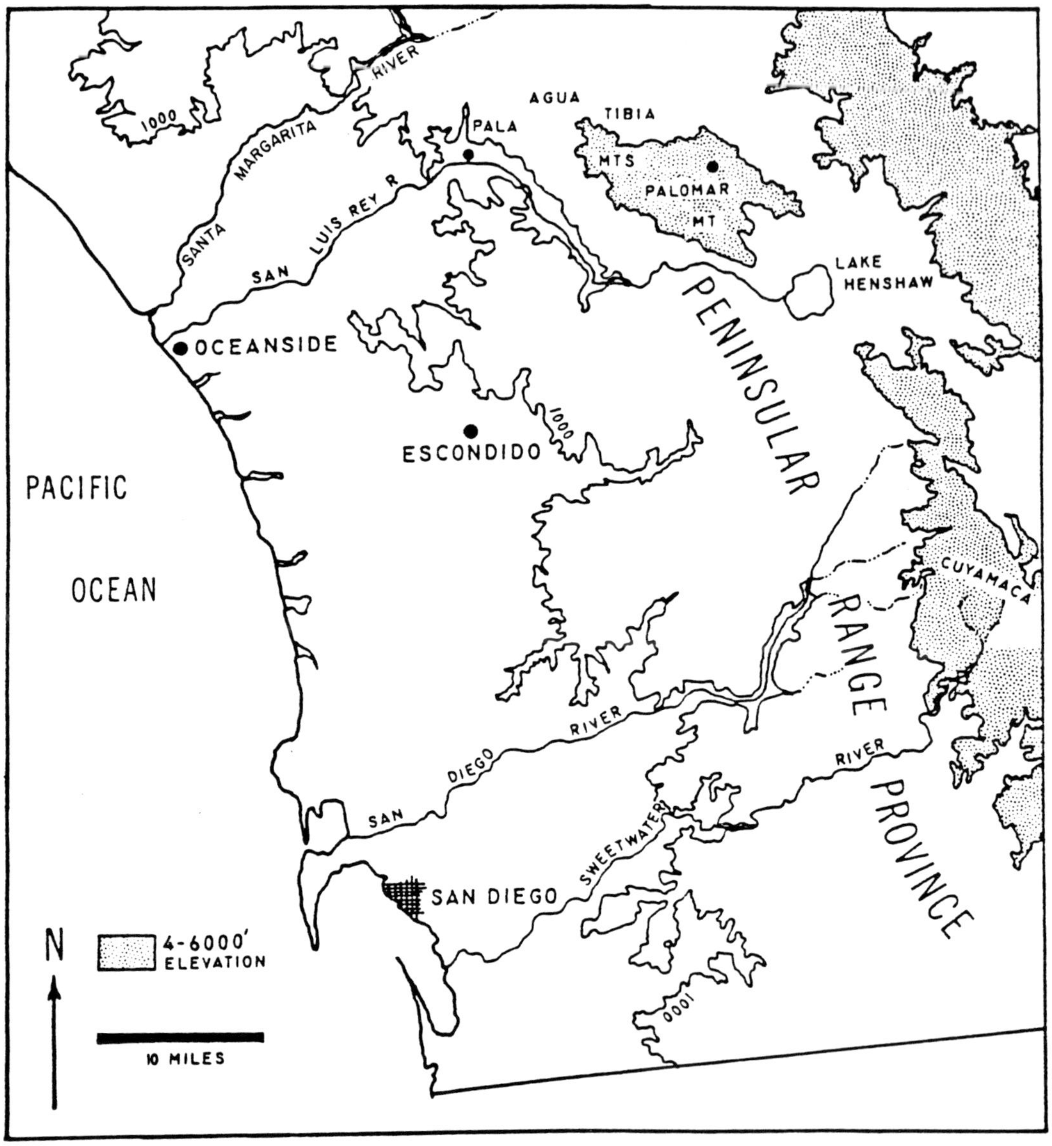

Figure 1. The southern California coastal region (from True et al. 1974).

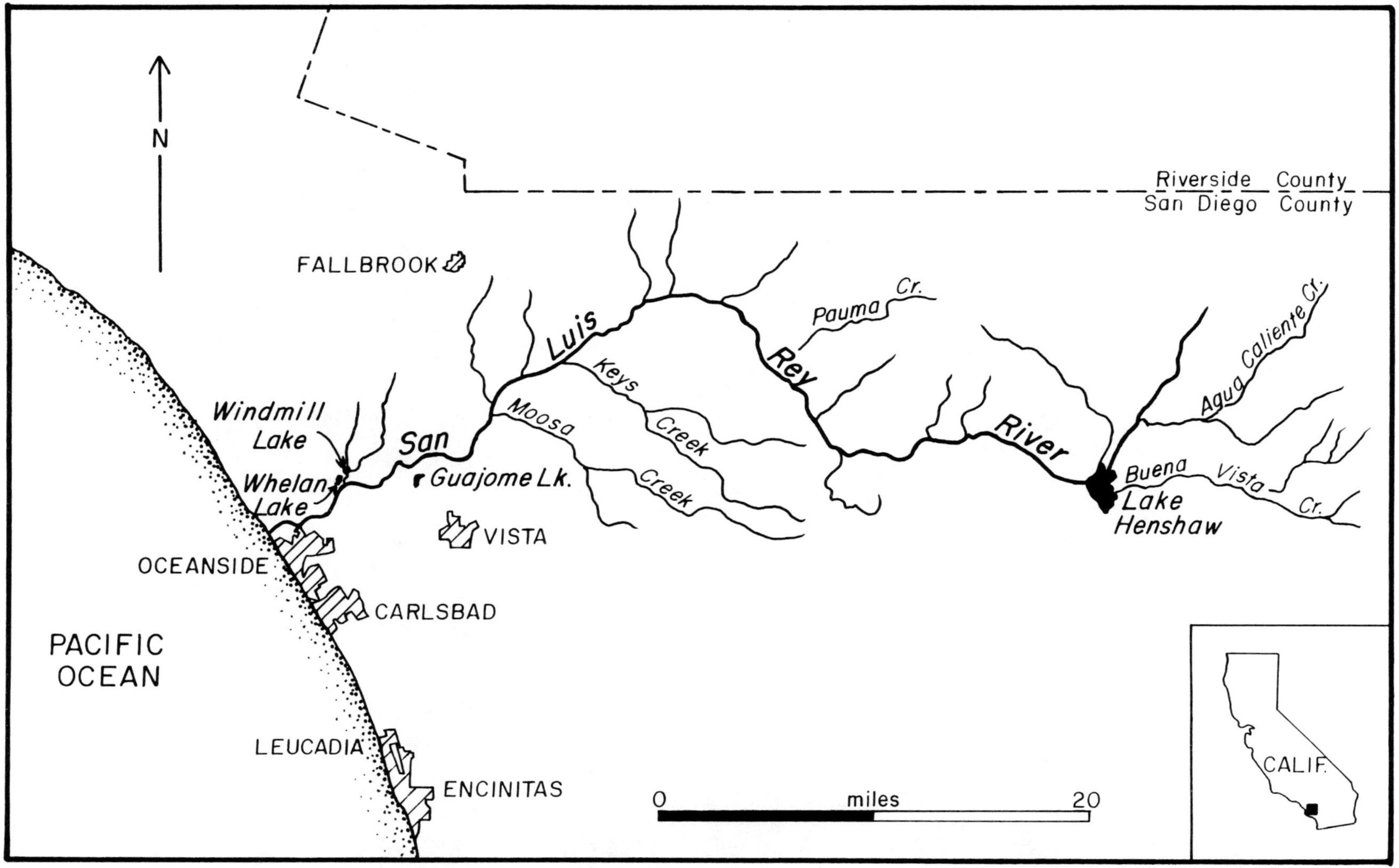

Figure 2. The San Luis Rey River Basin.

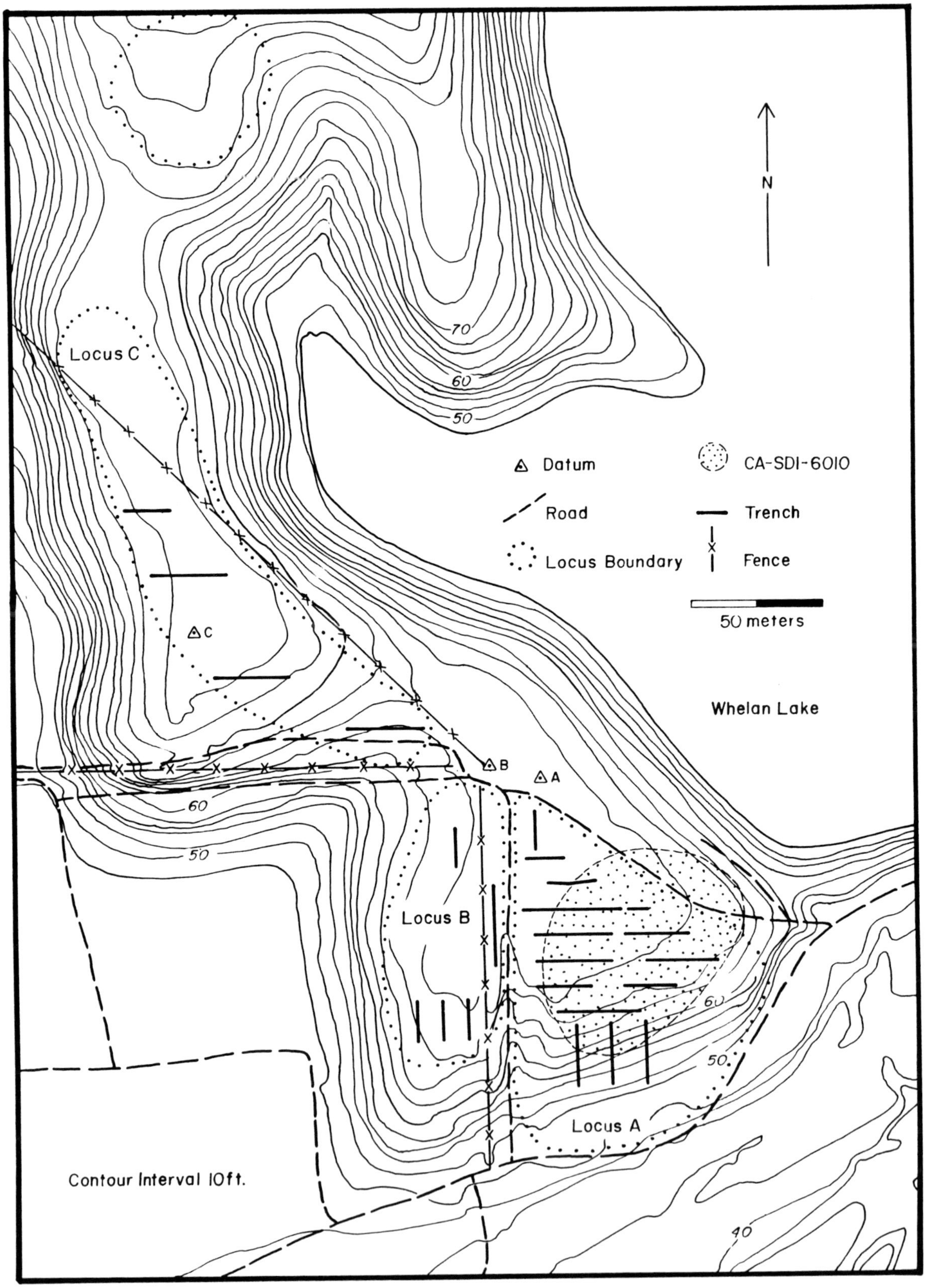

Figure 3. Map of the project area, showing the Whelan Lake site and test loci.

REPORT ORGANIZATION

The following report is organized in four parts. Part 1, comprised of Chapters 2 and 3, presents background information on the environment, culture history, and previous archaeological research in the region. It concludes in Chapter 4 with the research design that guided our work. Part 2 consists of field and laboratory methods and results. Chapter 5 presents the results of the fieldwork, whereas Chapters 6 through 10 provide the results and interpretations of various analytic studies. Part 3 presents our interpretations and recommendations. Chapter 11 provides a model of prehistoric adaptation to the San Luis Rey River basin, focusing on the La Jolla and Pauma cultures. The report concludes in Chapter 12 with a summary of our results and recommendations for future research and the management of the site.

CHAPTER 2
ENVIRONMENTAL BACKGROUND

INTRODUCTION

The project area is in the northern portion of the City of Oceanside, San Diego County, California. Site SDI-6010 is located immediately north of the San Luis Rey River and is situated on a knoll west of Whelan Lake. The San Luis Rey River drains steep uplands and wide terraces as it winds its way to the Pacific Ocean, approximately 7 km to the west. Camp Pendleton is a short distance to the north. The site is located in the gently sloping topography of the Coastal Plains of San Diego County, where the elevation on site ranges from 50 feet above mean sea level (AMSL) along the southern edge, to approximately 70 feet AMSL on the northern boundary (see Figure 3). The site has been transected in its southern portion by Whelan Road (an extension of North River Road).

The abundance of water and associated plants and animals made the San Luis Rey River basin one of the most attractive areas in northern San Diego County for settlement during prehistoric and ethnohistoric times. The basin attracted prehistoric Native Americans of the La Jollan and San Luis Rey cultures, who were followed by the historically known Luiseno. Later, the basin attracted the Spanish, who quickly established the Mission San Luis Rey (one mile southeast of the project area). In the 1830s, after the Mission period, the lush basin became home to many Hispanic ranchers and farmers. The basin is currently in the process of being intensively developed. The project area is presently covered with grasses and other weedy vegetation, and has recently served as grazing land for cattle and horses.

The source of the San Luis Rey River is above Lake Henshaw in the Agua Tibia Mountains. The river drains a number of smaller streams and intermittent creeks as it flows toward the San Luis Rey Lagoon and out to the coast. However, since the construction of the dam at Lake Henshaw and a number of gravel pits and associated check dams, the flow of water has been controlled. In fact, the purpose of the present project is to evaluate a site located on land that is proposed as a borrow pit for the construction of a flood control dam, one part of a larger COE project on the lower San Luis Rey River. Between its source and the ocean, the river flows through a number of plant communities, including coastal oak woodland, chaparral, coastal sage scrub, and fresh and salt water marshes (California Coastal Commission 1987:334). Plant associations near the site include oak woodland/woodland grass community with coast live oak (*Quercus agrifolia*) and associated undergrowth. Grazing and farming activities, however, have removed most of the native vegetation from the site.

In addition to the extensive modifications along the San Luis Rey River, a number of other major development and construction projects have altered the landscape in the surrounding area. Although not shown on the 1975 photorevised USGS, 7.5 minute Quadrangle, a large portion of the surrounding hillsides have been graded and developed with residential structures. Immediately east of Whelan Lake is a large dairy farm and just east of the farm is the San Luis Rey Wastewater Treatment Plant and aeration ponds. The treatment plant and dairy farm have significantly impacted Whelan Lake to the point where the lake now "totally functions as a sump storage basin" (Jehl 1989:2). These, as well as numerous other development projects, have significantly altered the composition of plants and animals in the San Luis Rey River basin.

To provide some background on the environmental setting of the area surrounding site SDI-6010, the geology, soils, climate, hydrology, flora, and fauna of the area are described in this chapter. The discussion of biotic resources highlights the historic alterations that have significantly altered the plants and animals of the region. In addition, geologic and hydrologic processes are discussed in order to provide the necessary data to evaluate site integrity and visibility in the region.

GEOLOGY

Southern California, as well as the entire west coast of North America, is extremely active geologically. Small-scale geologic processes such as earthquakes and volcanic events and broad-scale tectonic events which have caused uplift and subsidence of large land masses have shaped California's landscape. These processes, combined with alluvial and aeolian deposition and erosion, and sea level fluctuations reflecting glacial advance and retreat, are responsible for the modern appearance of the coastline. Approximately 150 million years ago, during the Jurassic Period, the California coast was situated eastward in the area now occupied by the Sierra Nevada (Caughman and Ginsberg 1981), and the current coastline location is where the Pacific Coast Geosyncline was positioned (Dunbar and Waage 1969). As the North American Plate shifted to the west, it overrode the Pacific Plate. Afterward, the California coast was extended westward over 100 miles and the Coast Ranges were created from sedimentary rock scraped from the Pacific Plate as it slid underneath the western boundary of the North American continent. The San Andreas fault zone marks the boundary between the Pacific and North American plates. The present configuration of the coast was essentially reached by about 6,000 or 7,000 B.P., a time corresponding to a rapid decline in sea level rise (see Bickel 1978:6 and Orme 1990:46).

The San Luis Rey project lies within the Peninsular Ranges Province, a large geomorphic province extending from the Los Angeles Basin southward to Baja California. The Peninsular Province consists of northwest-southeast trending mountains composed mainly of Mesozoic igneous and metamorphic rocks and Cenozoic sedimentary rocks. Younger materials in the Peninsular Province are primarily marine and terrestrial sedimentary rocks ranging in age from the Cretaceous Period (80 mya) to the Pleistocene Epoch (less than 2 mya) of the Quaternary Period (Sharp 1972).

In the lower San Luis Rey drainage, the Peninsular Range is characterized by irregular hills, which form the transition from the inland faulted ranges to the coastal plain (Sharp 1972). The uplands consist primarily of Cretaceous granodiorites intruded into older schists and quartzites between 90 and 100 mya (Campbell and Oakeshott 1966). The project area is underlain by fill soils, natural surficial soils, Quaternary alluvial soils, and Pleistocene river terrace deposits. At depth below these soils is sedimentary strata of the Tertiary Santiago Formation, which locally consists of dense sands interbedded with hard clay. Extensive zones of well-cemented sandstone are also known to exist in the area (Woodward-Clyde Consultants 1990). Figure 4 shows the geologic formations around SDI-6010.

SOILS

A surficial geological survey was conducted by Woodward-Clyde Consultants in 1990 for Affinis's environmental impact study of the Whelan Property (including portions of site SDI-6010). This data, and independent field analysis by SRI indicates that two types of soils occur on the site.

The southern portion of the site, below Whelan Road, consists of alluvium and slopewash. These soils are composed of loose to medium dense silty and clayey sands. The northern portion of the site lies on river terrace deposits. These deposits consist of "massive, medium dense to dense clayey sand soils containing some zones of sandy clay" (Affinis 1991:27).

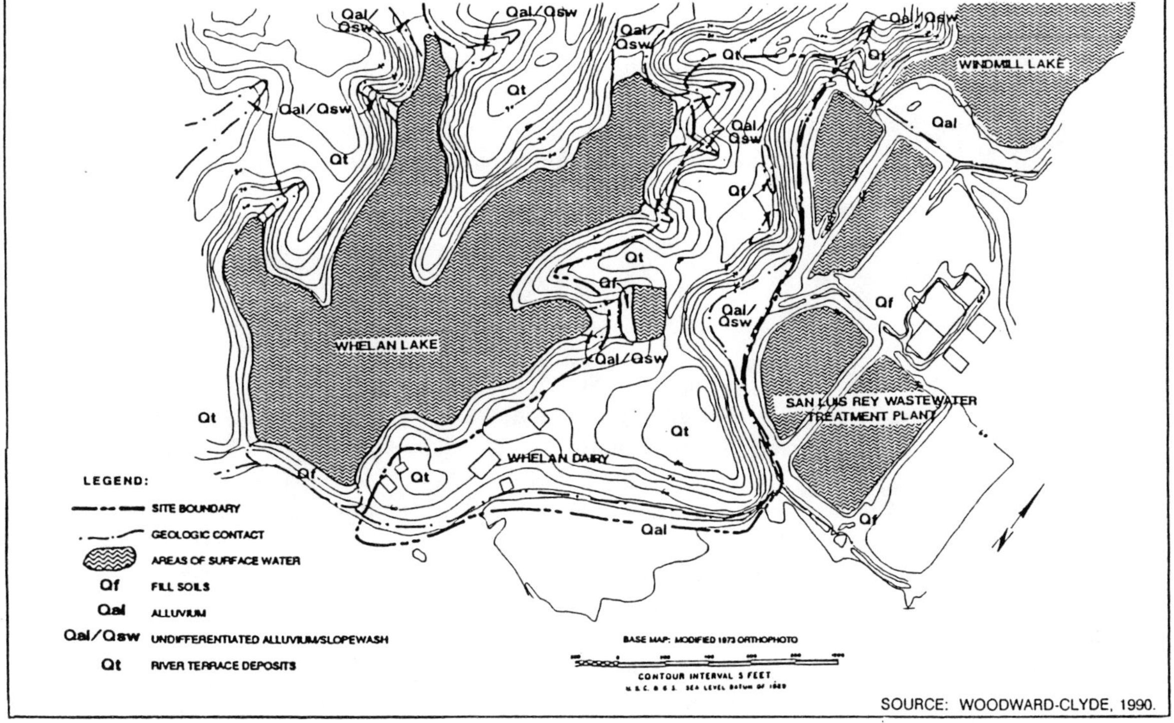

Figure 4. Local geologic formations (adapted from Woodward-Clyde Consultants 1990).

CLIMATE

California's coastal climate is classified as Mediterranean, that is, one characterized by two seasons, a temperate wet winter and a moderate dry summer. Only about one percent of the world has this type of climate, and it does not occur anywhere else in the United States (Caughman and Ginsberg 1981). Most precipitation falls between October and March, with the majority concentrated in December, January, and February. Rain virtually never falls during the summer months. The coastal climate is primarily influenced by the strong high pressure system located west of California and the presence of the ocean itself. This strong, high pressure system, known as the Pacific High, effectively blocks cool, moist breezes originating in the northern Pacific Ocean from moving onshore. The Pacific High is typically between 30° and 40° North Latitude during the summer months. The Pacific High weakens and moves south as winter approaches, allowing storms to occasionally reach land. High pressure systems occasionally form inland and cause a reversal of the prevailing northwesterly winds. These winds, called the Santa Anas sometimes lead to the rapid spread of wild fires.

Annual rainfall in the Oceanside area varies from 9 inches (23 cm) to 16 inches (41cm) and averages about 10 inches, with most falling in the cool winter months (Griner and Pryde 1976:37). An important property of the precipitation pattern is the overall unpredictability from one year to the next. The annual average rainfall is greatly exceeded during some years, with destructive flooding as a result. Conversely, some years only a small fraction of the average amount of rainfall occurs, and disasterous droughts sometimes result.

Due to the moderating effect of the Pacific Ocean, there is relatively little seasonal and daily variation in temperature along the entire California coast. Temperatures rarely exceed 100° F

(37.8° C) or fall below freezing in San Diego. Coastal areas in southern California average less than five days a year with maximum temperatures above 90° F (32.2° C) and no days below 32° F (0° C) (Fay et al. 1987). The mean minimum temperature in January, the coolest month, is 40° F (4.4° C) at the northwest corner of California, whereas San Diego, at 42° F (5.6° C), is only slightly warmer despite its location at the opposite end of the state (Beck and Haase 1974). Mean temperature in the Oceanside area is 59.2° F (Griner and Pryde 1976). Marine air keeps the relative humidity moderately high year round. Fog frequently occurs in the spring and summer months, but it generally "burns off" by noon.

HYDROLOGY

The San Luis Rey River has, until historic times, provided a reliable water supply to the entire basin. Seasonal water sources are also available throughout the basin, in the form of small streams flowing from side canyons. The overall area around SDI-6010 is in the Pilgrim Creek drainage basin (McIntire Group 1990). Whelan Lake, immediately east of the project area, occupies a small, lateral drainage of Pilgrim Creek and is formed by a small earthen dam at the drainage's southern end.

While the San Luis Rey River does provide a perennial source of water, it is subject to periods of severe flooding. Flood intervals center on 100 and 500 years. Site SDI-6010 is outside of the 500 year flood plain (according to 1987 FEMA maps), but the area below 58 feet elevation is within the 100 year flood plain of Pilgrim Creek and the San Luis Rey River (McIntire Group 1990). The 58-foot-elevation contour transects the southern portion of the site just north and upslope of Whelan Road.

In recent times, the construction of the dam at Lake Henshaw has served to control the flow of water in the San Luis Rey River. Also, recent flood control on the lower San Luis Rey River by the COE has served to check the flow of water. Pilgrim Creek has been channelized around the eastern and southern boundaries of the adjacent Whelan property. A small man-made pond exists to the east of Whelan Lake, and Windmill Lake is a short distance to the northeast.

VEGETATION

The lush vegetation along the San Luis Rey River and the coastal oak woodland community provided a wide variety of plant resources to aboriginal inhabitants. Farming, ranching, and development related to the construction of buildings, dams, roads, and other farm-related structures are responsible for removing virtually all native vegetation that once existed on the project area property and surrounding area. During prehistoric times, the coastal sage scrub community and grassland would most likely have covered the project area. Riparian and freshwater marsh communities would have been accessible along the San Luis Rey River, just south of the site. Affinis (1991) conducted a biological resources survey of the Whelan property immediately east of the project area and documented the existence of the wetland species presented in Table 1, and the upland vegetation species presented in Table 2.

Table 1. List of Wetland Vegetation Types Near Whelan Lake (after Affinis 1991:Appendix C:13-15).

Scientific Name	Common Name	Habitat**		
		FM	AM	RW
*Arundo donax**	Giant Reed			x
Azolla sp.	Pacific Mosquitofern	x		
Baccharix glutinosa	Mulefat			x
Carex sp.	Sedge	x		
Cirsium sp.*	Thistle			x
*Cotula coronopifolia**	Brass Buttons	x		
Cressa truxillensis	Alkali Weed		x	
*Cynodon dactylon**	Bermudagrass	x		x
Distichlis spicata	Coast Saltgrass		x	
Frankenia salina	Alkali Heath		x	
Heliotropium curassivicum	Heliotropium		x	
Isocoma veneta vernonioides	Coast Goldenbush	x	x	x
Lemna sp.	Duckweed	x		
*Nasturtium officinale**	White Water Cress	x		
*Paspalum dilatum**	Knotgrass	x		x
Plantago cf. *lanceolata**	Plantain	x		
Pluchea ororata	Salt Marsh Fleabane	x	x	x
Polygonum sp.	Knotweed	x		
*Polypogon monspeliensis**	Rabbitfoot Grass	x		x
Populus fremontii	Fremont Cottonwood			x
*Ricinus communis**	Castor Bean	x		
*Rumex crispus**	Curly Dock	x		
Salicornia cf. *virginica*	Pickleweed		x	
Salix laevigata	Red Willow			x
Salix goodingii	Black Willow			x
Salix lasiolepis	Arroyo Willow			x
Sambucus mexicana	Southwestern Elderberry			x
Scirpus cf. *olneyi*	Bulrush	x		
Sesuvium verrucosum	Sea Purslane		x	
Stachys sp.	Nettle	x		
*Tamarix parviflora**	Tamarisk			x
Typha cf. *latifolia*	Tall Cattail	x		
*Xanthium spinosum**	Spiny Cotbur	x		

* = Denotes introduced taxa
** FM = Freshwater Marsh, AM = Alkaline Marsh, RW = Riparian Woodland

Table 2. List of Upland Vegetation Types Near Whelan Lake (after Affinis 1991:Appendix C:16-17).

Scientific Name	Common Name	Vegetation Type**				
		D/R	IP	AG	CSS	NPG
Agrostis sp.*	Bentgrass		x			
Amaranthus sp.*	Tumbleweed	x				
Ambrosia psilostachys	Western Ragweed	x				
Artemisia californica	California Sagebrush			x	x	
Asclepias cf. *fascicularis*	Narrow-leaved Milkweed			x		
*Atriplex semibaccata**	Australian Saltbush	x				
*Atriplex patula hastata**	Spearscale	x				
Avena sp.*	Wild Oat	x	x			
Brassica sp.*	Mustard	x	x	x	x	x
*Bromus mollis**	Soft Chess	x		x	x	x
*Bromus rubens**	Red Brome	x		x	x	x
*Bromus diandrus**	Ripgut Grass	x		x	x	x
Chenopodium sp.*	Goosefoot	x				
*Cynodon dactylon**	Bermudagrass	x	x			
Encelia californica	Coast Encelia				x	
Erodium sp.*	Storksbill	x		x	x	x
*Foeniculum vulgare**	Sweet Fennel	x		x		
Hemizonia fasciculata	Fascicled Tarweed			x	x	x
Heterotheca sp.	Telegraph Weed	x				
Isocoma veneta vernonioides	Coast Goldenbush	x		x	x	x
Isomeris arborea	Bladderpod				x	x
Lolium sp.*	Wild Rye	x	x	x		
*Nicotiana glauca**	Tree Tobacco	x				
Opuntia cf. *prolifera*	Coast Cholla				x	x
Opuntia occidentalis	Thicket Tuna			x	x	
*Ricinis communis**	Castor Bean	x				
*Salsola australis**	Russian Thistle	x				
Stipa pulchra	Purple Needle Grass				x	x
*Xanthium spinosum**	Spiny Clotbur	x				

* = Denotes introduced taxa

** D/R = Disturbed/Ruderal, IP = Irrigated Pasture, AG = Annual Grassland,

CSS = Coastal Sage Scrub, NPG = Native Perennial Grassland

These studies identified 33 total wetland species with 12 introduced taxa and 29 total upland species with 18 introduced taxa. While nearly half of the species listed were introduced to the area, a number of these species existed on or near the site prior to disturbance. A full list of species in the area is beyond the scope of this project, but the locally available species important to the aboriginal population are listed below. In 1984, Quillen et al. compiled a partial list of the diverse local plants and their known ethnobotanic uses in the San Luis Rey area. Included among the many documented aboriginal uses are basketry, food, fiber, fuel, construction, and medicine. The list of species and their aboriginal use is reproduced below (Table 3).

Table 3. Paleoethnobotany of the San Luis Rey Area (adapted from Quillen et al. 1984).

Scientific Name	Common Name	Aboriginal Use
Artemisia californica	Sagebrush	Medicinal
Artemisia douglasiana	Mugwort	Arrow shafts
Claytonia perfoliata	Miner's Lettuce	Food
Eriodictyon crassifolium	Yerba Santa	Medicinal, soap, tea
Eriogonum fasciculatum	California Buckwheat	Medicinal
Heteromeles arbutifolia	Toyon	Beverage
Juncus sp.	Rush	Basketry
Platanus racemosa	Western Sycamore	Medicinal
Pluchea sericea	Arrow Weed	Arrow Shafts
Prunus ilicifolia	Holly-Leaved Cherry	Food, beverage
Quercus agrifolia	Coast Live Oak	Food, fuel
Quercus dumosa	Scrub Oak	Food, fuel, medicinal, basketry
Quercus englemannii	Englemann Oak	Food, fuel
Rhus laurina	Laurel Sumac	Basketry, twine
Rhus ovata	Sugar Bush	Beverages
Rhus trilobata	Squaw Bush	Basketry, twine, medicinal
Salix gooddingii	Black Willow	Basketry, construction, twine, bows
Salvia apiana	White Sage	Food, medicinal
Salvia mellifera	Black Sage	Food, medicinal
Sambucus mexicana	Elderberry	Medicinal, beverage, food, dye, construction
Toxicodendron diversilobum	Poison Oak	Medicinal
Xyloccus bicolor	Mission Manzanita	Beverage, food, tools, fuel
Yucca whipplei	Our Lord's Candle	Food, fiber, basketry

FAUNA

The San Luis Rey River basin continues to host a wide variety of fauna. Mammals, birds, reptiles, and invertebrates are found in several biotic communities. Availability of the larger mammals has, however, unquestionably declined since historic times due to farming, ranching, and the introduction of domesticated animals such as the horse and cow. Other animals, such as birds and invertebrates, may have changed little over time. While a modern list of terrestrial mammals present in the project area would serve little purpose, due to the farming and ranching activities on site, a recently gathered list of avifauna (Affinis 1991) and a list of archaeologically represented species at the nearby sites of SDI-5130 and SDI-6014 (Gibson et al. 1990) are provided to serve as a guide to some of the prehistorically available fauna.

Avifauna

A modern species list of avifauna (Affinis 1991), gathered during the middle of October 1990 in the Whelan Lake area, helps to establish the range of species locally available. This list of observed avifauna is displayed below, in Table 4.

Table 4. Observed Avifauna on the Whelan Property (adapted from Affinis 1991).

Scientific Name	Common Name
Podiceps nigricollis	Eared Grebe
Podilymbus podiceps	Pied-Billed Grebe
Phalacrocorax auritus	Double-Crested Cormorant
Ardea herodias	Great Blue Heron
Casmerodius albus	Great Egret
Egretta thula	Snowy Egret
Bubulcus ibis	Cattle Egret
Branta canadensis	Canada Goose
Butorides triatus	Green-Backed Heron
Nycticorax nycticorax	Black-Crowned Night-Heron
Anas platyrhynchos	Mallard
Anas acuta	Northern Pintail
Anas cyanoptera	Cinnamon Teal
Anas discors	Blue-Winged Teal
Anas streptera	Gadwall
Anas clypeata	Northern Shoveler
Anas americana	American Wigeon
Aythya collaris	Ring-Necked Duck
Oxyura jamaicensis	Ruddy Duck
Cathartes aura	Turkey Vulture
Elanus leucurus	Black-Shouldered Kite
Accipiter cooperi	Cooper's Hawk
Falco columbarius	Merlin
Buteo lineatus	Red-Shouldered Hawk
Buteo jamaicensis	Red-Tailed Hawk
Falco sparverius	American Kestrel
Gallinula chloropus	Common Moorhen
Fulica americana	American Coot
Himantopus mexicanus	Black-Necked Stilt
Charadrius vociferus	Killdeer
Actitis macularia	Spotted Sandpiper
Calidris minutilla	Least Sandpiper
Limnodromus griseus	Long-Billed Dowitcher
Larus delawarensis	Ring-Billed Gull
Larus californicus	California Gull
Sterna forsteri	Forster's Tern
Columba livia	Rock Dove
Zenaida macroura	Mourning Dove

Table 4. Observed Avifauna on the Whelan Property (adapted from Affinis 1991). Continued.

Scientific Name	Common Name
Chaetura vauxi	Vaux's Swift
Calypte anna	Anna's Hummingbird
Picoides nuttallii	Nuttall's Woodpecker
Picoides pubescens	Downy Woodpecker
Megaceryle alcyon	Belted Kingfisher
Tyrannus vociferans	Cassin's Kingbird
Sayornis nigricans	Black Phoebe
Sayornis saya	Say's Phoebe
Tachycineta bicolor	Tree Swallow
Hirunda pyrrhonota	Cliff Swallow
Hirundo rustica	Barn Swallow
Corvus corax	Common Raven
Psaltriparus minimus	Bushtit
Troglodytes aedon	House Wren
Thryomanes bewickii	Bewick's Wren
Telmatodytes palustris	Marsh Wren
Mimus polyglottus	Northern Mockingbird
Lanius ludovicianus	Loggerhead Shrike
Sturnus vulgaris	Eurasian Starling
Vermivora celata	Orange-Crowned Warbler
Dendroica petechia	Yellow Warbler
Dendroica coronata	Yellow-Rumped Warbler
Dendroica townsendi	Townsend's Warbler
Geothlypis trichas	Common Yellowthroat
Passerculus sandwichensis	Savannah Sparrow
Chondestes grammacus	Lark Sparrow
Melospiza lincolnii	Lincoln's Sparrow
Melospiza melodia	Song Sparrow
Zonotrichia leucophrys	White-Crowned Sparrow
Agelaius phoeniceus	Red-Winged Blackbird
Agelaius tricolor	Tricolored Blackbird
Sturnella neglecta	Western Meadowlark
Euphagus cyanocephalus	Brewer's Blackbird
Cassidix mexicanus	Great-Tailed Grackle
Molothus ater	Brown-Headed Cowbird
Carpodacus mexicanus	House Finch
Carduelis psaltria	Lesser Goldfinch
Carduelis tristis	American Goldfinch
Passer domesticus	House Sparrow

While 76 total species were observed in the immediate area, there is little doubt that a number of these have been attracted by the recent construction of the San Luis Rey Wastewater Treatment Plant and associated ponds and marshes. The environment in the immediate area is, however, conducive to all of the birds listed. In addition, the list of observed avifauna was constructed during the Fall, and the list clearly reflects seasonal patterns. Prehistoric environmental differences would probably not significantly change the list, since the lagoon is still present a short distance downstream. For a partial list of prehistoric fauna in the San Luis Rey River area refer to Table 5.

Fauna In Prehistory

A sample of faunal remains was excavated from SDI-5130 and SDI-6014 by INFOTEC Research, Inc. and Greenwood and Associates in 1990 and analyzed by the Zooarchaeological Laboratory of the Institute of Archaeology at University of California at Los Angeles (UCLA). Because these sites are only a short distance downstream from the current project area, it is reasonable to assume that the faunal remains are indicative of prehistoric subsistence patterns practiced in the project area. Table 5, constructed below, lists all of the identified faunal remains at these two sites.

Table 5. Identified Faunal Remains from SDI-5130 and SDI-6014 (adapted from Gibson et al. 1990).

Scientific Name	Common Name
Thomomys bottae	Gopher
Lepus californicus	Black-Tailed Hare
Sylvilagus audubonii	Audubon Cottontail
Odocoileus hemionus	Mule Deer
Sus scrofa	Pig
Canis sp.	Dog/Coyote
Amnospermophilus beecheyi	California Ground Squirrel
Equus equus, Equus sp.	Horse/Burro
Bos taurus	Cow
Microtus californicus	California Meadow Mouse
Ovis aries	Sheep, Domestic
Dipodomys merriami	Merriam Kangeroo Rat
Mephitis mephitis	Striped Skunk
Procyon lotor	Raccoon
Perognathus sp.	Pocket Mouse
Sciurus griseus	Western Gray Squirrel
Felis domesticus	Cat
Anas sp.	Duck
Lophortyx californicus	California Quail
Gallus gallus	Chicken
Clemmys marmorata	Pond Turtle
Coluber constrictor	Black Racer Snake
Pituophis catenifer	Gopher Snake
Thamnophis sirtalis	Garter Snake
Mylobatis californica	Bat Ray
Rhinobatus productus	Guitarfish (Shoveler's Shark)
Sebastes sp.	Rockfish
Triakis semifasciata	Leopard Shark
Prionace glauca	Blue Shark
Thunnus sp.	Tuna

Table 5. Identified Faunal Remains from SDI-5130 and SDI-6014 (adapted from Gibson et al. 1990). Continued.

Scientific Name	Common Name
Embiotucidae	Surf Perch
Semicossypus pulchur	Sheephead
Sardinops sajax	Sardine
Leureathes tennis	Grunion
Paralabrax clathratus	Kelp Bass
Sarda chiliensis	Pacific Bonito
Cancer sp.	Crab
Lucina nassula	Woven Lucine
Tivela stultorum	Pismo Clam
Siliqua sp.	Razor Clam
Saxidomus nuttali	Common Washington Clam
Serpulorbis squamice	Corroding Worm Shell
Chama sp.	Jewel Box
Macoma sp.	Macoma
Polinices lewisii	Lewis' Moon Shell
Septifer bifurcatus	Bifurcate Mussel
Tellina sp.	Tellin

A number of the identified species (e.g., cow, horse, pig) are domesticates, indicating a historic period of occupation in the upper levels of the sites. All of the identified species, with the possible exceptions of gophers and mice, were considered to be used for food. In addition, all of the identified species can still be found locally.

CONCLUSIONS

A number of conclusions regarding human habitation of the San Luis Rey River basin can be drawn from the environmental background. First, the area is more conducive to a diversified subsistence economy than it is to a specialized one. From the headwaters of the river to the lagoon, a number of microenvironments are encountered and each supports a diverse range of flora and fauna. No one resource or set of resources is so abundant that a successful adaptation could depend exclusively on them. While it could be argued that the inland Pauma Culture was adapted to acorn exploitation, ethnographically documented population movements clearly indicate that this reliance was seasonal. In addition, as discussed below in Chapter 3, coastal resources could have been significant to the earliest inhabitants of California (see Jones 1991). The San Luis Rey Lagoon would have been an ideal location to gather marine resources and, when compared to a reliance on terrestrial resources, it seems highly probable that the earliest inhabitants of this area would have exploited the coast.

While many areas of southern California are faced with a limited amount of water, this is not the case in the San Luis Rey River basin. Agriculture and horticulture practices were relatively simple and required only minor water conservation practices. As documented by the missionaries, the abundant resources and ample water supply made the river basin an ideal location for the Mission. These same factors probably drew prehistoric populations to the area. The mild climate would not have dictated heavy clothing or serious shelters, but winter rains and summer aridity encouraged seasonal population movements.

Finally, the archaeological record in the lower San Luis Rey River basin is a biased representation of prehistoric settlement patterns. First, rising sea levels could have inundated a number of early sites along the coast, biasing dates of occupation in favor of later periods. Second, because the area is subject to 100 and 500 year floods, most of the early sites along the river would most likely have been destroyed or buried by flood waters and sediment. As a result, flood episodes bias the discovery of sites to the higher river terraces and mountain tops. Thus, sites will be visible on the surface primarily on older, more stable, land surfaces. Caution needs to be taken when interpreting local settlement patterns. In addition, models that account for only surface remains are likely to be inadequate as tools for explaining the past in the coastal area.

A GEOMORPHOLOGICAL ASSESSMENT OF THE STUDY AREA

Antony R. Orme

The Whelan Lake site lies just above the northern edge of the floodplain of the San Luis Rey River, 7 km northeast of the river's outlet to the Pacific Ocean at Oceanside, California. The site is located at an elevation of approximately 20 m above sea level on a low knoll on a late Pleistocene river terrace immediately southeast of Whelan Lake. From the geomorphological and pedological perspective, the location of the site is unexceptional. It lies safely above the maximum flood limits of the historic floods of 1916 and 1938, and was presumably favored because of its easy access to a variety of habitats: the main flood channel with its lingering seasonal fresh water which, prior to the creation of Lake Henshaw, may have been nearly perennial; the adjacent marshes along the main channel; the marshes and seasonal pond on the site of modern Whelan Lake; the chaparral-covered coastal mesas; and the coast and nearby estuaries with their rich resources. The site is also somewhat sheltered by hills from the prevailing north-westerly winds.

An understanding of the physical character of the site, including its geomorphology and sedimentology, requires an appreciation of its regional setting. Accordingly, this study first examines the San Luis Rey River basin and then the area adjacent to the site that includes the river floodplain and terrace country around Whelan Lake. An appraisal of the geomorphic characteristics of the site itself follows in Chapter 5.

Regional Setting: The San Luis Rey River Basin

The Drainage Basin

The San Luis Rey River drains some 1450 km^2 of the Peninsular Ranges in northern San Diego County. Rising just below the crest of Bucksnort Mountain (1888 m above sea level), the river flows 25 km southwest into the Lake Henshaw basin where it is joined by several other headwater tributaries including its West Fork which rises below Mount Palomar (1872 m) (see Figure 2). From Lake Henshaw, the river runs generally northwest for 38 km along the margins of the Elsinore Fault zone to Pala where it turns southwest for a further 37 km to the ocean. The main stream channel is thus 100 km long and has a mean gradient of 19 m/km, although the latter is much steeper above and below the relatively gentle Lake Henshaw basin.

Lake Henshaw is the only large storage reservoir in the San Luis Rey River Basin. Completed in 1922, the reservoir controls a drainage area of 531 km^2 and has a capacity of 240 x 10^6 m^3. Water from the lake is released both down the mainstream and diverted some 15 km downstream into the Escondido Mutual Water Company canal to Wohlford Lake in the Escondido Creek basin. This canal has functioned since 1896. Any interpretation of the flood history of the San Luis Rey River near Whelan Lake must therefore recognize that, to some extent since 1896 and particularly since 1922, downstream flooding has been partly mitigated by upstream control. Before 1922, flood discharges and sediment delivery past the Whelan Lake site would have been significantly larger that in later years. A number of small reservoirs have also been constructed at various times near the mouths of minor tributaries, most notably Guajome Lake, Windmill Lake and Whelan Lake in the lower part of the basin.

Relief within the basin varies from moderate to rugged with a maximum elevation of 1950 m and a mean basin elevation of about 700 m. In general, mean annual precipitation increases with elevation - from 25 cm/yr along the coast to 55-65 cm/yr at Henshaw Dam and Mount Palomar Observatory, to perhaps 100 cm/yr on the highest ridge crests (National Oceanic and Atmospheric Administration 1990). Coniferous forest occurs on the higher ridges and oak woodland at intermediate elevations, but chaparral dominates the remainder of the basin. Approximately 30% of the basin has been developed to a greater or lesser extent for agriculture, including citrus groves and avocado orchards, and a further 10% of the basin is urbanized, notably inland from Oceanside.

Geologic Framework

For much of its length, the San Luis Rey River and its tributaries drain an extensive area of Cretaceous granitic rocks forming the complex southern California batholith. These rocks comprise mostly granodiorite, tonalite, diorite, and gabbro, with numerous pegmatite dikes and extensive mineralization. Associated with these batholithic rocks, usually as roof pendants, are various prebatholithic metasediments and metavolcanics, notably schists, gneisses, quartzites and marbles. About 15 km from the coast, the prebatholithic and batholithic rocks are covered by Tertiary sediments, mostly Eocene and Miocene marine conglomerates, sandstones, siltstones, and shales, which in turn are veneered with Quaternary marine-terrace and beach-ridge deposits. Interior valleys are characterized by Quaternary alluvial and colluvial sediments derived from older rocks, notably around Lake Henshaw, while the main river channel downstream from Rincon and Pala contains alluvial fill up to 60 m deep that readily favors water storage.

The southern California batholith has been raised as a block, severely fractured, and tilted gently seawards by Cenozoic tectonism. It has been further dislocated by extensive strike-slip faulting associated in the middle and upper San Luis Rey River basin with the NW-SE striking Elsinore Fault zone, a 20-km-wide zone of faulting that includes the Agua Tibia and Agua Caliente faults. This combination of coarse crystalline rocks, uplift, fracturing, and faulting has favored deep chemical weathering and disintegration which have in turn ensured a ready supply of rock waste to the drainage system. Lithologies such as those noted above are the principal components of modern alluvium and terrace deposits in the vicinity of Whelan Lake.

Hydrology and Flood History

The hydrologic behavior and flood history of the San Luis Rey River reflect both the underlying geological framework and the magnitude and frequency of storm rainfall. In the middle and upper parts of the basin underlain by deeply weathered, coarse crystalline rocks and relatively porous granular soils, light to moderate rains occurring early in the rainy season or after prolonged drought will be largely absorbed by the regolith and any water that reaches the streams will likely percolate into the alluvial fill beneath the channels. On the less permeable Tertiary sediments and indurated Quaternary deposits closer to the coast, some run-off is likely to follow all but the lightest storms. Heavy and intense rainfall later in the rainy season, especially if preceded by other heavy rains, is likely to produce flood flows from the mountain areas and such flows will probably not be reduced by infiltration into channel fills.

Channel gradient will also affect the hydrologic response. Although the average gradient of the San Luis Rey River is 19 m/km, the uppermost reaches and the middle reach below Lake Henshaw approximate 50 to 60 m/km. The lower reaches downstream from Pala have an average gradient of only 3 m/km, although the south-bank tributaries of Key Canyon and Canyon Creeks have gradients of 20 m/km and 15 m/km respectively. Even before the construction of Lake Henshaw in 1922, storm run-off from the upper basin would have decelerated and deposited a significant portion of its sediment load in the gentler terrain of the Elsinore Fault zone, as shown by the widespread accumulation of Quaternary alluvium around the present reservoir. The steeper gradient downstream from the lake, together with the relatively steep tributary gradients, would, however, have ensured a resurgence of flood activity in the middle and lower reaches of the river. The decreased gradient below Pala would in turn promote widespread deposition, particularly of the coarser fraction of the sediment load. This would seem to account for the thick alluvial fills and broad floodplain development in the vicinity of Mission San Luis Rey and Whelan Lake, further favored by the higher present base level imposed after the culmination of the main Flandrian transgression some 5000 years ago.

Some 32 gauging stations have operated within the San Luis Rey catchment at one time or another but only 9 of these have records exceeding 20 years. The most important from the perspective of the Whelan Lake environment is the USGS station located 1.8 km upstream from the river mouth in Oceanside. This station, which has operated intermittently since 1912, is the only one within the basin to record both stream discharge and sediment data. In order to calculate natural annual flows at Oceanside, for the purposes of estimating prehistoric flows, it is necessary to add the observed or actual annual flow at Oceanside to the net annual flow into the Escondido Mutual Water Company canal as adjusted for percolation and waste. Data for the period 1930-75 indicate a range of natural annual flows from 0.27×10^6 m^3 in 1961 to 169.97×10^6 m^3 in 1938 and 222.78×10^6 m^3 in 1937 (Brownlie et al. 1979). In other words, in a flood year the total discharge may approach 1000 times the discharge of a relatively dry year, and the two years noted above may not represent the extremes to be found over a longer, prehistoric record. Using similar adjustments, the natural annual suspended sediment yield during the 1930-75 period was found to average 191,000 tonnes but to range from 72 tonnes in 1961 to 2,680,000 tonnes by adding 10% for estimated bedload. Mean daily flows and sediment yields show similar erratic behavior over the course of a single year.

The floods of December 1937 and March 1938 were indeed spectacular and affected most of coastal southern California. The latter flood, which was associated with a series of rainstorms between February 27 and March 4, was undoubtedly enhanced by high antecedent moisture attributable to several prior storms beginning the previous December. Indeed, whereas the March 1938 event appears to have been the largest flood over the past 100 years, the mean areal rainfall over the San Luis Rey river basin for the storm period was only 28.5 cm, compared with 57.4 cm during the storm period of January 1916.

Southern California's flood history may be extended back to 1770 by means of mission records, court transcripts and other documentary and oral evidence (Lynch 1931; Troxell 1942), but the flood behavior of the San Luis Rey River is not well known. Certainly, there were major floods in San Diego County in 1821, 1861-62, and 1883-84 (when Escondido recorded 128 cm of rainfall during the year), and floods were recorded on the San Luis Rey River above Bonsall in 1891. Measurements of flood discharge were first taken in 1889 and since that time the record is clearer. During the present century, apart from the major events of 1937-38, significant floods occurred in 1916, 1941, 1969, 1978, and 1980. The February 1980 flood caused significant erosion along the San Luis Rey River between Rincon and Pala and also in Moosa Canyon, as well as extensive flooding between Bonsall and Oceanside, notably across the floodplain near Mission San Luis Rey.

The prime reason for providing this background information of the San Luis Rey River basin is to place some of the observations and conclusions regarding the Whelan Lake area and the archaeological site in their setting.

The Whelan Lake Area

The concept of the morphostratigraphic unit is now introduced to characterize the terrain in the vicinity of Whelan Lake. A morphostratigraphic unit is a terrain component distinguished by the association of a specific landform with the underlying geology. Figure 5 recognizes five morphostratigraphic units around Whelan Lake.

Unit 1, dissected Tertiary bedrock, characterizes the hilly country and sideslopes above the Lower San Luis Rey River. Most of the area is underlain by Eocene marine sandstones, siltstones, and shales, overlain by Miocene marine siltstones and shales within 4 km of the coast. These rocks have been well dissected by fluvial and hillslope processes, with flat land narrowly confined to the interfluves which rise to around 100 m above sea level north and south of Whelan Lake.

Unit 2, the Pleistocene marine-terrace terrain, veneers the Miocene rocks towards the coast. The terraces rise from low bluffs along the coast and river mouth to form relatively extensive mesas up to 100 m above sea level, the innermost of which have been much dissected by tributary erosion. The terraces in turn carry a number of sub-parallel beach ridges representing stillstands in the overall regression of middle and late Pleistocene seas. Fragments of marine terraces occur at higher elevations beyond the study area. The sequence ranges in age from approximately 500,000 to 100,000 years B.P., and provides an upper age limit for the later morphostratigraphic units.

Unit 3 comprises a series of late Pleistocene river terraces which, within the study area depicted in Figure 5, rise to 30-35 m above sea level. These terraces represent fragments of former floodplains that developed as the San Luis Rey River extended itself seawards across the emergent marine terraces during late Pleistocene times, probably within the past 100,000 years. These river terraces, which are underlain at variable depth by Eocene marine sediments, are characterized by poorly sorted to moderately sorted fluvial deposits containing abundant clasts of granodiorite, tonalite, diorite, gabbro, and various metasedimentary and metavolcanic rocks in a coarse sandy matrix. The small valleys that are now occupied by Whelan Lake are cut into this terrace unit.

The present floodplain of the San Luis Rey River is represented by Unit 4, the Holocene alluvium. These fluvial deposits accumulated during successive floods during late prehistoric and historic times. Within 12 km of the sea, the floodplain gradient reduces to 2 m/km, compared with the average river gradient of 19 m/km and a steep section of 50-60 m/km below Lake Henshaw. This reduction in gradient reduces stream velocity and therefore sediment-transporting competence and capacity. The most recent major rise of sea level, the Flandrian transgression which culminated around 5000 years ago, further reduced this gradient and promoted the extensive deposition of thick alluvial deposits around Mission San Luis Rey. Even before recent alluviation, however, it is unlikely that the sea, in the form of a restricted estuary in the Oceanside Narrows, extended more than 2 km inland from the coast. At a point 2 km from the river mouth, the present floodplain is no more than 180 m wide and if one removes the alluvium which has accumulated to 3-5 m above sea level, the floodplain becomes little more than 150 m wide. In such a narrow restriction, a seawater lagoon, even if it

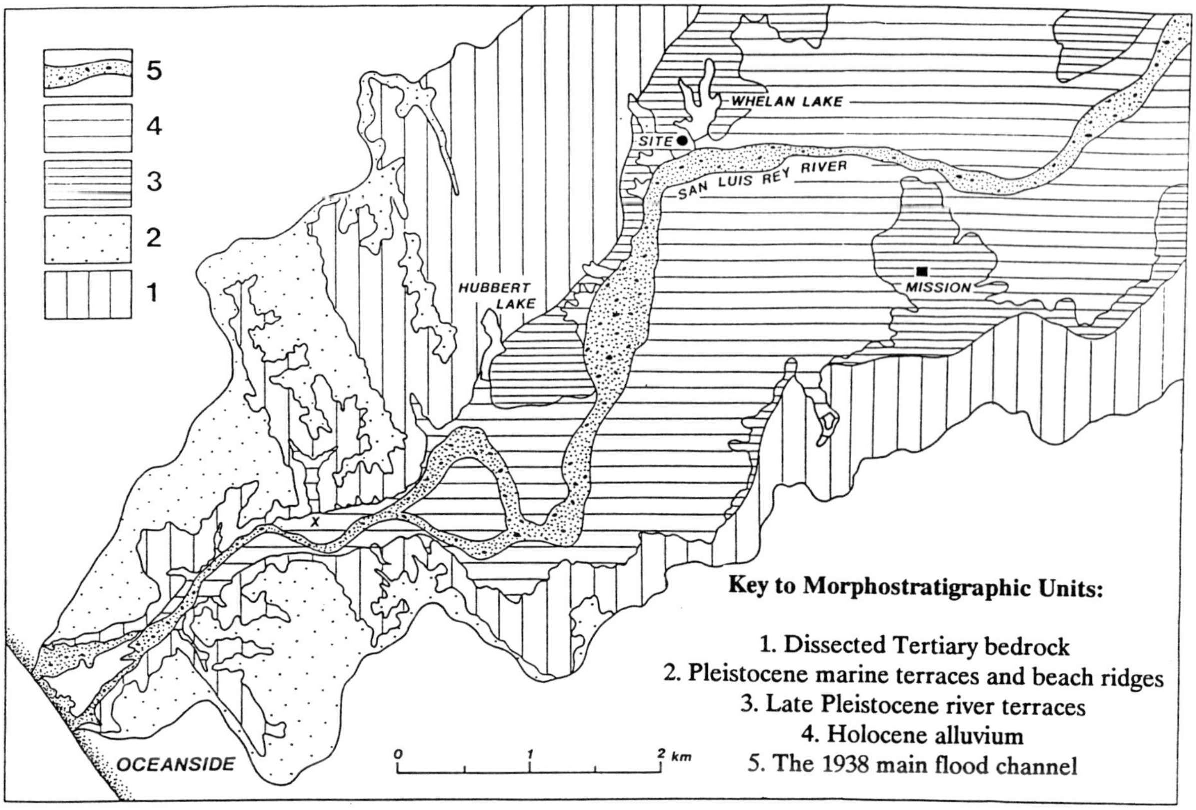

Figure 5. The lower San Luis Rey River basin showing morphostratigraphic units (by Antony Orme).

penetrated farther upstream, would surely become choked by tidal and fluvial sediment in a very short time, except during major floods when the lagoon, its sediments and fauna, would be flushed seawards. Thus, even if a lagoon did exist around Whelan Lake, it would soon become brackish and eventually freshwater, certainly within a few hundred years after seawater reached its present level.

Freshwater marshes would have formed farther upstream as waters backed up behind the Oceanside Narrows estuary. Historic maps, notably the USGS editions of 1898 and 1901, show extensive marshes in the floodplain above the Narrows, with the old coach road crossing the floodplain just east of these marshes from San Luis Rey to the terrace west of Whelan Lake. Thus, a freshwater lagoon and marsh existed in the Whelan Lake vicinity (between the north and south banks of the San Luis Rey River floodplain), but not a saltwater lagoon. This was subsequently filled with alluvial sediment, probably quite quickly but we have no knowledge of sediment rates locally.

Unit 5 represents the main channel of the major channel of the flood of March 1938 and has been reconstructed from field and aerial photographic evidence. Owing to subsequent dike construction along this flood course, this channel does not differ significantly from the present channel, except in the vicinity of Oceanside Airport. There, 1938 floodwaters backed up behind the Oceanside Narrows and spread widely across the floodplain. An artificial cut through the neck of the southward-projecting meander, accompanied by nearby dykes, subsequently redirected the river towards its present course.

It is appropriate at this time to examine the origins of Whelan Lake. In 1869, part of the San Luis Rey River basin was surveyed by James Pascoe who, in his field notes, mentions a "small lagoon" and "Pio Pico's field" are clearly depicted on Pascoe's map of 1870. The lagoon again appears in a sketch of

1881 that appeared in Elliot's (1883) *History of San Diego County*, and in the USGS map of 1898. Ariotti (n.d.) later notes that "In 1946, Ellen Whelan had a small dam erected at the south end of the lake which increased the lake size to about twenty-five acres and enabled Miss Whelan to irrigate the pasture on the west side. Later, in 1958, the City of Oceanside, in agreement with Miss Whelan, increased the height of the dam, and the lake size to thirty-seven acres. In addition, the city built a road twenty feet wide across the dam." The field evidence fully supports these two episodes of dam construction but Ariotti's statement and earlier narrative imply that the small dam erected in 1946 was perhaps the first such structure across the "small lagoon". This has in turn led to the belief, mentioned to this investigator, that Whelan Lake was a natural feature augmented by later construction.

Careful evaluation of field evidence and aerial photography indicates to the investigator that, as a perennial water body, Whelan Lake is not a natural feature but an artificial creation that occupies the site of a small marsh and perhaps shallow seasonal pond that accumulated behind sidewash deposits of the San Luis Rey River. There are several examples of similar "lakes" in the area, notably Guajome Lake and Hubbert Lake, that have been created by the artificial damming of small tributaries. However, the best analogy occurs at point X in Figure 5, 4 km downstream of Whelan Lake, where the mouth flood sediments some time prior to 1938, thereby creating a natural marsh. The nature of this deposition is clearly evident on the 1938 aerial photographs where a wedge of sediment slopes away from the mainstream into the tributary valley. During major floods, the main channel carries a large sediment load and on deposition the channel bed is commonly raised above the floor of neighboring tributaries, creating temporary ponds and later marshes in the latter. Such sites would appeal to early farmers and graziers as localities where, with but modest raising of the natural dam, a more permanent body of water could be trapped and stored. Because a "small lagoon" already existed by 1869, it seems likely that raising of the natural dam occurred during the preceding 71 years -- sometime after the founding of Mission San Luis Rey in 1798. Prior to 1798, the natural marsh and seasonal pond would presumably have attracted wildfowl, but Whelan Lake in its form of the period 1869-1946 most probably did not exist.

CHAPTER 3
CURRENT ISSUES IN CULTURE HISTORY

INTRODUCTION

The purpose of this chapter is to provide the necessary background for the construction of an appropriate research design for the Whelan Lake site. The chapter is organized in four sections. The first section provides a brief review of the important archaeological investigations in the project region. This section provides the reader with information concerning the projects that have contributed to the process of defining the particular cultural sequences employed in southern California. Next, this chapter discusses the chronology that has been established over the course of archaeological investigations. This section, highlighting chronology, is presented in order to explain the often confusing terminology used in the many proposed cultural sequences. Following the section on chronological issues is a critical examination of the current problematic issues of cultural distinction and subsistence in prehistory. Finally, the protohistoric and historic periods are examined. While we rely on archaeological investigations to examine prehistory, ethnographic and ethnohistoric accounts of the project area help to focus on the problems surrounding the issue of culture and chronology in the historic period.

Since this review is not exhaustive, the reader is referred to Moratto (1984), and Chartkoff and Chartkoff (1984) for general information on the archaeology and culture history of the southern California coastal region. The reader is referred to Bean and Shipek (1978), Kroeber (1925), and White (1963), for excellent ethnographic and ethnohistoric overviews of the Luiseno. For specific information on particular excavations the reader is referred to the individual site reports referenced below. A regional map (Figure 6) provides the locations of the sites and areas covered by surveys in the lower San Luis Rey River basin, discussed below.

PREVIOUS ARCHAEOLOGICAL INVESTIGATIONS

Since the early archaeological investigations in the San Diego area (Rogers 1929a, 1939), archaeologists have generally focused their attention on issues concerning chronology, economic systems (subsistence and technology), and settlement patterns. During the past few decades, the amount of investigative projects has increased due to both private and public development projects spawned by the population increase in southern California. While this increase in archaeological projects has not completely resolved any of the problematic issues, the cumulative effect of the investigations has increased our knowledge and understanding of the San Diego region. Below, the historical development and current status of these major issues are discussed. This section is divided according to Carrico's (1987) temporal periods of archaeological investigation in southern California (i.e., 1920 to 1955, 1955 to 1972, and 1973 to present).

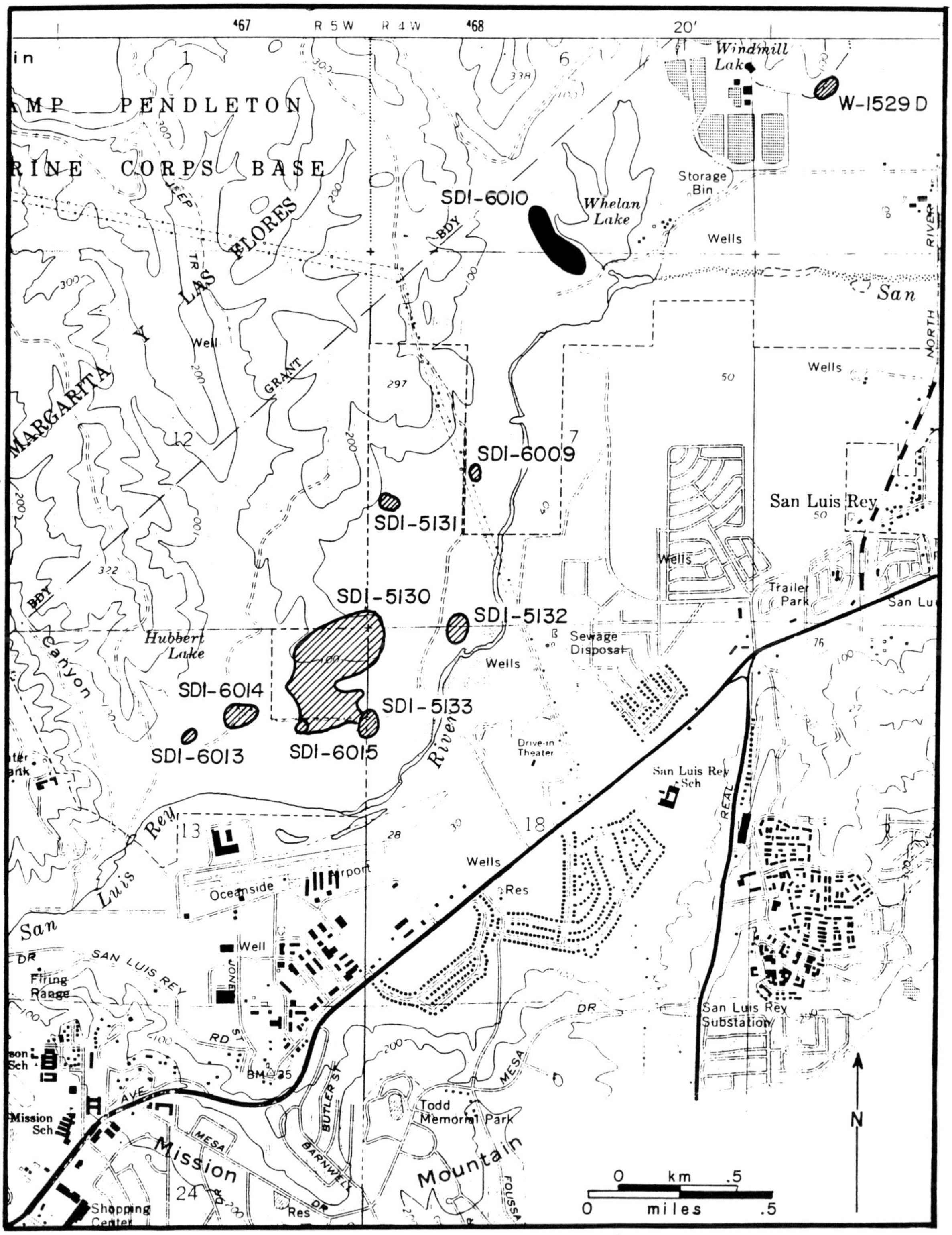

Figure 6. Regional map showing site locations.

Early Archaeological Investigations 1920 - 1955

While pothunters and collectors have been excavating in California since the late 1800s, the first archaeological investigations took place in the 1920s (Moratto 1984). In the 1920s, in the San Dieguito River Channel, M.J. Rogers located, recorded, and wrote about the culture he labeled the "Scraper Maker People." Excavations, however, did not begin until 1936. Rogers described his findings in a brief report (1938), discussed related events in a later article (1939), and finally published a culture history in 1945.

Shortly after Rogers' work, Carter (1949, 1950, 1951, 1952, 1954, 1957), in a series of articles, began to claim that Early Man was in the New World as early as 20,000 to 30,000 years B.P. Carter suggested that the bluffs in La Jolla held information that proved Early Man was in the San Diego area. This work, while ignored by many, came to serve as an important source for later Early Man advocates (Carrico 1987).

In the early 1930s, some limited excavations took place at the Mission San Luis Rey (Soto 1961). These excavations focused on architectural features and generally overlooked small cultural deposits. Excavations in the 1950s supplemented the early excavations and revealed bathing facilities and other water supply features. A large number of prehistoric and historic artifacts were recovered during these excavations (Soto 1961). Unfortunately, little information exists from the early excavations at the mission, and only very brief analyses were conducted during the later excavations.

Most of the investigations during this early period of southern California archaeology focused on issues concerning chronology. Attempts to define cultures based on artifact assemblages and subsistence practices began, and the first chronological sequence for the region was established (Rogers 1945). This early period laid the foundation for future investigations and started the controversy surrounding chronological issues.

The Middle Years 1955 - 1972

Rogers continued to work in the San Diego region into the 1960s, although during the later period he was no longer alone. During the 1950s the area began to be investigated by an increasing number of individuals from the University of California, Los Angeles (UCLA). With the passing of Rogers, work in San Diego County was then almost exclusively done by the archaeologists from UCLA. Based on surface surveys of the San Luis Rey River basin, Meighan (1954) defined the San Luis Rey Culture. In that same year, True (1954) described pictographs in the San Luis Rey River basin. In 1955, Wallace provided a chronology for the area. Closing out the 1950s, True (1958), described the Pauma Culture, which fit into neither Rogers' (1945) nor Wallace's (1955) earlier chronologies. The surveys conducted during these years began to question the existing chronological sequences proposed for the area.

In the early 1960s, Claude Warren began his work in the coastal San Diego area. Adding to Rogers' (1945) earlier work, Warren and others (Warren and True 1961; Warren et al. 1961) conducted surveys of river basins in western San Diego County. Also in 1961, excavation reports of a La Jollan Culture site at the Scripps Estates in La Jolla were published (Shumway et al. 1961). In 1963, excavations at Batiquitos Lagoon (Crabtree et al. 1963) provided archaeologists with a wealth of information concerning shell middens and associated artifacts (Carrico 1987). These investigations added a great deal of information to the existing chronology and began to fill in the voids of the transition periods.

In 1964, Warren completed his doctoral dissertation that synthesized cultural change and continuity on the southern California coast. Warren's dissertation was complemented by Pourade's (1966) edited volume, *Ancient Hunters of the Far West*, that updated Rogers' earlier ideas. In 1966, True completed his dissertation which focused on the archaeological variation between Yuman and Shoshonean groups in San Diego County. True (1970, 1977) continued his investigations on the Cuyamaca complex into the 1970s and further defined its characteristics. Later in the 1960s, Hayden (1966), Warren (1967), and Moriarty (1967), attempted to clear up the misconceptions of the San Dieguito Culture and the transition to the La Jollan Culture (discussed in detail below). While archaeological projects from the academic world (such as those just discussed) have not stopped, they have, in recent years, been far outnumbered by projects conducted under contract. The relevant contract projects are discussed below.

Recent Archaeological Investigations 1973 - Present

In 1973, contract archaeology began to take over the bulk of archaeological excavations in San Diego County, as well as in many other areas in the United States. Since this time, spurred by population growth and concomitant development, hundreds of cultural resource surveys and environmental impact studies have blanketed the southern California region. Since a complete review of these studies would be nearly impossible, only those studies relevant to the present project will be reviewed. The reader is directed to the individual site reports for in-depth discussions of material and data analyses.

While a number of projects were conducted in the San Luis Rey region between 1973 and 1977, all were conducted outside of the lower basin. The number of archaeological projects in the basin increased significantly in the late 1970s, however, due to the COE's flood control project and other private developments. Several of these projects have been in the general area of SDI-6010. In 1977, six prehistoric deposits were discovered on the Whelan Ranch Survey (RECON 1977) and given one site number, W-1529 (A-F). Also in 1977, Christopher Drover, working independently, recorded and described an additional four sites. SDI-5130 (W-1361) was described as a widespread area containing many artifacts, shell, and exotic lithics (Drover 1977). SDI-5131 (W-1362) was an historic site with the remains of a number of structures and privy depressions. SDI-5132 (W-1363) was described as a localized area of exotic lithics and shell fragments, later determined to be slope wash from the adjacent site, SDI-5130 (Carrico 1978). SDI-5133 (W-1364) was recorded as an historic site consisting of several adobe structures and a well.

In 1978, WESTEC personnel conducted a survey in the lower San Luis Rey River drainage and discovered eight sites (Franklin and Carrico 1978). Excluding the present SDI-6010 (W-1834), all of these sites are downstream from the project area. SDI-6008 (W-1832) was located directly in the river drainage. Fragments of Tizon Brown ware, along with minimal numbers of lithics, were observed on the surface in the original survey. Upon return one year later, however, the site was found to have been flood damaged and a subsurface test proved negative (Franklin and Carrico 1979). SDI-6009 (W-1833) contained both historic and prehistoric material and subsurface testing yielded cultural material, consisting of bone, shell, ceramics, and lithics, to a depth in excess of 120 cm. SDI-6011 (W-1835) was highly impacted by river flooding, and subsurface testing revealed a disrupted vertical and horizontal context. The location of SDI-6012 (W-1836) and the sparseness of cultural material led the investigators to the conclusion that the material was redeposited during flood stages from another site upstream.

Whereas WESTEC personnel recorded SDI-6010 in 1978, they did not conduct a data recovery program (surface or subsurface collection) because the site was outside of their project area. The site was originally described as follows: "At this spot a moderately dense concentration of shellfish (*Chione californiensis, Chione undatella,* and *Aquipecten acquisulcatus*), hammerpounders, scrapers and flakes of basalt material constituting the lithic assemblage were found" (Franklin and Carrico 1978:28-29). Based on surface observations of the field crew, the site measured 80 meters east-west, by 60 meters north-south. In 1988, WESTEC tested an artifact scatter thought to be an extension of SDI-6010. The testing included 10 backhoe trenches and one 1 m x 1 m unit. The excavations led the investigators to conclude that the midden and artifacts were brought to this location (possibly from SDI-6010) and redeposited (Kyle et al. 1988:4-17). Artifacts found included one scraper, one hammerstone, two manos, a fragment of worked glass, some lithic debitage, and shell.

In 1978, WESTEC mapped SDI-5130 and conducted surface and subsurface testing comprised of eight 1 m x 1 m units (Carrico 1978). Excavations determined that the upper 40 cm of the site had been disturbed by plowing and discing. It was concluded that the area investigated was probably a small section of a large La Jolla III or early San Luis Rey period campsite. In 1984, WESTEC completed the mitigation of the impact area of the site by excavating an additional 30 1 m x 1 m units (Quillen et al. 1984). Artifacts recovered from these excavations included milling tools, flaked tools, lithic debitage, a steatite disc bead, shell beads, and an Elko Corner-notched projectile point. In 1988, WESTEC conducted additional excavations at the site for the Oceanside Annexation Project (Kyle et al. 1988). Excavations included two 1 m x 1 m units and 25 backhoe trenches along the northern edge of the site. Based on the results of these investigations and radiocarbon dates indicating occupation from ca. 5,000 to 6,000 years B.P., the site was determined to be a major La Jollan village. The northern edge of the site was interpreted as a lithic reduction area that contained some hearths.

In 1989, Cultural Resource Management Services (CRMS) returned to SDI-5130, as well as other sites, to develop a testing plan for that portion of the site that would be affected by COE actions in regards to proposed flood control projects (Cultural Resource Management Services 1989). It was concluded based on auger testing that SDI-6015 was a separate site outside of the boundaries of SDI-5130. Also it was discovered that SDI-5133 had been completely covered by overburden. Based on these investigations, CRMS (1989) and Wells and Farrell (1990b) interpreted SDI-5130 as having two smaller sites entirely contained within its boundaries, SDI-5133 (an historic component), and SDI-6015 (a locus of prehistoric material).

Tying into previous grid systems, Gibson et al. (1990) excavated 45 1 m x 1 m units and 38 backhoe trenches as part of test excavations at SDI-5130. Although the integrity of the site was impaired by both natural and cultural processes, many intact features were found to exist. Four burials and a number of hearths were discovered. Employing both relative and absolute dating techniques, two major episodes of occupation were found to exist, the first being the La Jolla complex (6,000 B.C. - A.D. 1), and the second being the late San Luis Rey complex (post-A.D. 1750), (Meighan 1954, 1959; Moriarty 1966; Moratto 1984; True et al. 1974; Warren 1964, 1968; True 1980). Based on this work, Gibson et al. (1990) reversed Wells and Farrell's conclusions and argued that SDI-6015 was in fact part of SDI-5130.

SDI-5133 was originally recorded as having several historic structures (Drover 1977), but by the time of the 1990 investigations (Gibson et al. 1990), these structures were not visible. Two historic features were discovered in subsurface testing, a scatter of floor tiles and a trash pit (with human remains). Two prehistoric features were also discovered in this area.

Further upstream from SDI-5130 and SDI-5133 are SDI-6013 and SDI-6014. SDI-6013 (W-1837) was recorded in 1978 by WESTEC personnel (Franklin and Carrico 1978) and was tested by WESTEC in 1979 (Franklin and Carrico 1979). Although the site was reported as limited in areal extent, the test unit was excavated to 110 cm depth (halted due to constraints on work space, not a lack of cultural material), reflecting a relatively deep deposit. The site was also investigated by Wells and Farrell (1990a, 1990b) who reported that the site had been used for dumping fill and that intact cultural material was not found. They suggested that the site consisted of redeposited material.

SDI-6014 (W-1838) was recorded by WESTEC personnel in 1978 (Franklin and Carrico 1978) and tested by WESTEC in 1979 (Franklin and Carrico 1979). Although surface collection was limited due to heavy vegetation, the excavated unit yielded three radiocarbon dates that suggested a San Luis Rey II period of occupation. Wells and Farrell (1990a, 1990b) tested SDI-6014 and recommended additional testing of the site with the exception of the western portion that is in the floodplain and does not warrant testing. Additional testing was conducted in 1990 (Gibson et al. 1990). Ten backhoe trenches, 40 sondage units, and three test pits yielded a very small assemblage of lithics, one projectile point, one composite tool, three ground stone fragments, and one human radius fragment. The upper 60 cm were severely disturbed and no datable material was found.

In 1989, Affinis conducted a cultural resource survey of the Whelan property as part of an overall environmental impact study for the City of Oceanside (Affinis 1991). During this survey two prehistoric sites (SDI-11,468, SDI-11,469) and one historic site (SDI-11,470) were recorded. SDI-11,468 was described as a light scatter of shellfish remains and lithic artifacts. The site has been subjected to a great deal of disturbance and a reliable interpretation of the contents was impossible. SDI-11,469 was identified as "a small milling station with 5 bedrock mortars and no artifacts" (Affinis 1991:71). SDI-11,470 was identified as an historic complex of "houses, barns, and outbuildings located at the southwestern edge of the dairy portion of the project area, along Whelan Lake" (Affinis 1991:72). These structures may have been built by the Pico family as early as the 1850s.

Historical archaeological investigations near Mission San Luis Rey have resulted in a much better understanding of the history of the area. The most recent work was the second phase of excavation conducted by Greenwood and Associates (1991) at SDI-5133H and SDI-6014 for the COE. This work resulted in the further documentation of the Matthew Hubbard homestead and the Preston Hubbard adobe, historical ranches of the late 1870s. Additional work conducted near Mission San Luis Rey (SDI-241) in 1990 by Magalousis and Kelsey (1990) resulted in new information regarding the Mission and surrounding area. Caltrans has recently conducted archaeological investigations, not yet published, along the San Luis Rey River, adding new information to the growing data base of the region (Judith Tordoff, personal communication 1991). The Guajome Rancho, located nearby, has also been the subject of a number of investigations (Engstrand and Ward n.d.; Fink 1974, 1980; Van Horn 1983; Noah 1986; Hector and Manley 1987, 1988). A site representing the interface between the mission and the Native American community was tested in 1978 (Greenwood 1978).

It is true that the number of investigations in this latest period of archaeological excavations has both added to the data base concerning culture change in the San Diego area and provided a great deal of information on small outlying sites not investigated in previous archaeological endeavors. Unfortunately, most of the projects conducted under contract have, due to constraints of the project areas and costs involved, focused on site-specific problems, overlooking regional chronological issues and settlement patterns. Thus, while many issues concerning intrasite settlement and subsistence patterns have been addressed, most of the broader problematic issues, outlined by earlier researchers, have been ignored. Although not inherently problematic, the nature of contract archaeology has changed the focus of investigations from regional concerns to site-specific issues.

In short, three major phases of archaeological investigations have taken place in the San Diego area. The first phase was from the 1920s to 1955. Most of the work conducted during the first phase was carried out by Rogers and Carter and focused on defining individual cultures and establishing the chronology of the region. The second phase was from 1955 to 1972, when UCLA conducted its investigations upstream from the current project area. These projects refined the earlier chronologies and began to fully describe the individual cultures in the area. Many problematic issues were, however, left unresolved. The third phase of investigations began in 1973, when contract archaeology appeared on the scene and began to address site-specific issues. In 1977, the COE began flood control projects on the lower San Luis Rey River. With the exception of recording sites, all COE-sponsored projects have been downstream of SDI-6010. SDI-6010, therefore, geographically falls between two previously investigated areas in the San Luis Rey River basin. As a result, this project serves to update regional settlement patterns and helps connect the two previously investigated areas. Most of the information outlined below is a result of the empirical studies discussed above. Each archaeological survey and excavation has added some data to aid in the reconstruction of the cultural sequence and past lifeways of the cultures in the southern California coastal area.

ESTABLISHING CHRONOLOGY

Establishing the chronology of the southern California region has been a difficult task for archaeologists for a number of reasons. First, the paucity of investigated and well-dated sites in the region has created a sequence based on a dispersed group of sites. Second, archaeologists and ethnographers have attempted to systematize and order a continuous, slow-moving cultural process. Third, variation in the timing and character of coastal resource use has led to a poor understanding of the foraging ability of the early California inhabitants. Finally, investigators have failed to agree on the defining characteristics of a culture, and as a result have created a confusing terminology. This section serves to highlight the problems facing archaeologists who attempt to establish a cultural chronology in a region where cultural boundaries are blurry. A complete discussion of material culture, subsistence, and settlement pattern differences between cultures is provided in the following section. We begin this section by tackling the terminology used in the region.

Terminology

"The demands on the language of archaeology are greater now than ever before..." (Reid 1991:388). This recent statement by the editor of *American Antiquity* shows that the concerns regarding terminology and language are widespread. In California, the terminology is confusing to both the public and the professional. In order to comprehend the proposed cultural sequence of southern California, one must understand the differences between a "tradition," a "complex," and a "culture." Archaeologists begin by defining assemblages of artifacts, features, and sites, within a specific time period, that share similar characteristics. An assemblage of material culture is described as a tradition when it "embraces the archaeology of a large part of [California] for a considerable period of time, and during that span of time the archaeology reflects continuity in social structure and adaptation even when details may vary from site to site" (Chartkoff and Chartkoff 1984:17). A number of other terms, including "complex", "phase", and "horizon" have been used to describe more limited spans of time and space. Unfortunately, the use of these terms is not consistent among archaeologists. As a result, the terms are often used interchangeably and are usually defined to fit specific reports or papers.

While use of the term "culture" goes back over a hundred years to the roots of anthropology, usage of the term in California can be traced back to Kroeber's concept of a "culture area." The California culture area (different from the modern political boundaries of the state) was divided, primarily by ethnographers, into a number of smaller culture areas based on an essentially arbitrary inventory of culture traits (see Figure 7 for a map of the California culture areas). Archaeologists borrowed this means of dividing larger culture areas and applied it to the archaeological record. Since every different chronology, constructed over the years, has employed its own terminology, the term "culture" will be used here for the sake of consistency and in order to relieve some of the confusion generated from the proliferation of terms. It is not our intent to create a new definition of the term or

Figure 7. Map of California culture areas (from Chartkoff and Chartkoff 1984).

even to solve the problems associated with the use of a multiplicity of terms, rather we hope to simply avoid the controversy and confusion. For example, whereas Meighan (1959) describes the La Jollan "complex", we will refer to this as the La Jollan "culture." "Tradition" is used here as a term that ties a number of cultures together across time and space, such as Warren's (1968) "Encinitas Tradition." Refer to Table 6 as a guide to the different terminologies used in cultural sequences over the years.

Table 6. Commonly Used Cultural Sequences in the San Diego area (from Moratto 1984).

Years	Coastal San Diego County (Rogers 1939, 1945)	Coastal San Diego County (Moriarty 1966)	Interior San Diego County, Northern (Meighan 1954; True 1958, 1966, 1970)	Interior San Diego County, Southern (Meighan 1954; True 1958, 1966, 1970)	Syntheses (Wallace 1955)	Syntheses (Meighan 1959)	Syntheses (Warren 1968)	Syntheses (Wallace 1978)
	Yuman III Culture		Luiseño	Diegueño	Horizon IV Late Prehistoric		Yuman / Shoshonean	
1500	Yuman II Culture		San Luis Rey II / San Luis Rey I	Cuyamaca Complex				
1000	Yuman I Culture		Shoshonean Intrusion					
500 A.D.					Horizon III Intermediate			
0 B.C.			Transition or Hiatus ?					
500								Period III: Diversified Subsistence
1000								
1500	La Jolla II Culture	La Jolla III				Archaic Stage		
2000							Encinitas Tradition	
2500		La Jolla II	Millingstone Substratum (La Jolla/Pauma Complexes)		Horizon II Millingstone			
3000								
3500	La Jolla I Culture							
4000								
4500		La Jolla I						Period II: Food Collecting
5000								
5500								
6000	San Dieguito Culture	San Dieguito	San Dieguito		Horizon I Early Man	Early Lithic Stage	San Dieguito Tradition	Period I: Hunting

The Sequence Begins

The cultural sequence of the San Diego County region, including the San Luis Rey River basin, is generally separated into three major cultures (complexes or traditions): San Dieguito, Encinitas (La Jolla-Pauma), and Late Prehistoric (Warren and True 1961; Rogers 1939; Harding 1951; Warren 1968; Wallace 1955; Moratto 1984; Gallegos and Hector 1987). While it has been suggested that humans were in the San Diego area during the Pleistocene (Carter 1949, 1957; Childers 1974; Childers and Minshall 1980), this evidence is controversial and generally unaccepted. This section outlines the accepted cultural sequence of the region (see Table 6). A critical examination of the similarities and differences between the cultures and a discussion of the ethnography of the Luiseno is provided in the next section.

Malcom Rogers (1929c) established the first chronology of the San Diego area in 1929. After severe flooding of the San Dieguito River, Rogers discovered an archaeological site exposed in the channel. Based on his observations at this site, Rogers later identified an early "Shell Midden People" and a later "Scraper Maker culture." Rogers fully investigated this site (SDI-149) in 1939 and, based on the stratigraphic data and artifacts recovered, reversed the chronological order of the cultures and renamed the earlier culture San Dieguito and the later culture La Jolla (Rogers 1939, 1945). Also in 1939, Rogers defined a sequence of industries in the lower Colorado River area: Malpais, Playa I, and Playa II. In 1958, Rogers recognized similarities in the coastal and Colorado River assemblages and renamed the entire sequence San Dieguito (I, II, and III). Later, the interior desert and lower Colorado River cultures were renamed the Lake Mojave culture (Wallace 1962, following Cambell et al. 1937). According to Wallace (1962), the Lake Mojave culture reflects early adaptations to the changing environment and is very similar to the San Dieguito culture. Soon after these changes, Warren (1967) offered his definition of the San Dieguito complex and included the interior desert units of Lake Mojave, Death Valley I, Playa I and II, and the coastal units, claiming that they were all regional variants of the same culture. Others (Donnan 1964; Wallace 1962, 1978) maintained that Lake Mojave was a distinct culture adapted to a different environment. While the regional sequence is still a topic of debate, a number of syntheses have been proposed.

Of the many sequences proposed for the southern California area (see Table 6), two main regional syntheses are commonly used in modern archaeological interpretations. The first was advanced by W.J. Wallace in 1955 and revised in 1978. In his original synthesis, Wallace (1955) defined four cultural horizons each with local variation: Early Man, Millingstone, Intermediate cultures, and Late Prehistoric cultures. When this synthesis was employed in the San Luis Rey region, the local variations that fell into each horizon were: Horizon I, San Dieguito; Horizon II, La Jolla I; Horizon III, La Jolla II; and Horizon IV, the San Luis Rey culture. Wallace's (1978) revision divided the sequence into three broader periods: Period I, hunting; Period II, food collecting; and Period III, diversified subsistence. Between Wallace's (1955) original synthesis and his revision (Wallace 1978), Warren (1968) advanced his regional synthesis. Employing a more ecological approach, Warren defined five traditions in southern California. In the San Diego area three of the five traditions are represented: (1) San Dieguito, (2) La Jolla (the regional representation of his Encinitas Tradition), and (3) San Luis Rey (the regional representation of his Campbell Tradition). In addition to the Colorado River and interior desert regions, the San Dieguito culture was later incorporated into the larger Western Pluvial Lakes Tradition that extended from northeastern California to the Mojave Desert and the San Diego coastal area (Bedwell 1970). Each proposed sequence differs in terminology and dates, and each has its advantages and disadvantages. One may employ any of the sequences, depending on whether one wishes to emphasize a culture's chronological order in a local region or in a broad area.

The proliferation of sequences over the years reflects a number of changes and developments in the field of archaeology in general (see Willey and Sabloff 1980). Early sequences (Rogers 1945) focus on subsistence and technology, reflecting archaeology's emphasis on classifying artifacts. As social relationships, ideology, and other aspects of culture began to be investigated in the 1950s, an emphasis on cultural ecology (see Meighan 1959c) began to appear in the chronology. Warren's (1968) synthesis reflects this ecological approach in archaeology. A major problem with all of the sequences is that they tend to separate cultures based on very slight differences, such as the La Jolla-Pauma distinction, which is primarily based on a difference in resource utilization. Similarly, the important cultural transition from hunting cultures to gathering cultures, is based on functional characteristics of their material culture. Identifying synchronic cultural differences and diachronic culture change, based on these distinctions, is a very imprecise procedure. The slight differences between the cultures reflects the fact that change in California has been a long and slow process. The problems associated with distinguishing between cultures are reflected in the number of cultural sequences advanced for the region. As a result, while other regions of the United States have focused on much broader cultural sequences, California continues to define and redefine its local schemes (see Chartkoff and Chartkoff 1984 for an attempt to incorporate the California sequence into a larger scheme). The cultural sequence of the southern California region is now outlined (an in-depth discussion of the settlement and subsistence patterns of these cultures follows this outline). While precise dates for the transitions are avoided, the cultures discussed are generally considered to follow the sequence outlined below.

Early Man?

While none of the chronologies discussed above include a culture prior to the San Dieguito, it is necessary to include a brief note on the debate that surrounds this topic. Some archaeologists (Leakey et al. 1968, 1972; Simpson 1980; Singer 1977; Schuiling 1979; Childers and Minshall 1980) claim that humans have been in the deserts of eastern California and coastal southern California in excess of 15,000 years. Unfortunately, most of the early bones, features, and artifacts from these sites have been dated using questionable methods or materials. The site that has gained the most attention is in the Calico Hills, on the shore of Pleistocene Lake Manix. Thousands of "artifacts" have been recovered from this site, but both the dates of the stones and the "cultural" nature of the stones have come into question. The techniques used to date the stones are questionable and Haynes (1969, 1973) claims that the stones are not tools but rather, simply geofacts. Other sites that have received some attention include the Yuha Pinto Wash (Childers and Minshall 1980), the Yuha Burial (Barker et al. 1973; Childers 1974; Bischoff et al. 1976, 1978), and the La Jolla Shores and Del Mar Sites (Bada et al. 1974). The problem remains however, that while Pleistocene surfaces do in fact exist, conclusive evidence reflecting early human habitation of these surfaces has not yet been found.

Hunting Cultures

Rogers originally discovered the San Dieguito type site SDI-149 (now generally known as the C.W. Harris site), in 1928, and started publishing on the culture shortly thereafter (Rogers 1929a, 1929b, 1929c, 1938, 1939, 1945). In 1937, George Carter discovered a cultural deposit in the same area and excavations started in the San Dieguito River Channel in 1938 (Warren 1966). Many years later, Warren and True (1961) continued work at SDI-149 and further defined the culture. This culture dates from circa 10,000 to 7,500 B.P. (Moratto 1984), and was adapted to the post-Pleistocene environment in which the megafauna had largely disappeared and the hotter, drier climate had forced groups to settle near reliable water sources. The San Dieguito culture, as defined by Warren (1967), was a hunting culture, with a flaked stone industry that included large flake and core scrapers, choppers, hammerstones, drills and gravers. While plant processing artifacts are virtually absent from the assemblage, there is little doubt that this culture used plant resources when available. Sites are primarily found in eastern San Diego County in ancient lake terraces and no sites of this early culture have been found in the vicinity of the current project area.

The San Dieguito culture has been divided into three phases (Rogers 1939; Wallace 1955). Relatively sparse evidence of the San Dieguito I phase indicates that the assemblages include ovate bifaces, scraper planes, and chopping tools. Sites from this phase of occupation are found in the inland desert regions. San Dieguito II sites are found in San Diego County and contain smaller projectile points, a greater variety of choppers and scrapers, and more finely worked lithics. Kaldenberg and Ezell (1974) and Gallegos and Carrico (1984) found that this culture was also at least partially dependent on marine resources. San Dieguito III assemblages display highly varied lithic forms. In addition to scrapers and choppers, the assemblage also includes pressure flaked blades and points, crescentic stones, and bifacial knives (Rogers 1939, 1945, 1966). The division of this culture into these distinct phases has, however, been challenged and criticized (see especially, Bull 1987; Ezell 1987). San Dieguito has also been lumped with a number of other cultures (Playa, Lake Mojave, Death Valley I) into the larger Western Pluvial Lakes Tradition (Bedwell 1970).

The Transition Period from Hunting to Food Collecting

Although the transition period from San Dieguito to La Jolla was unexplained by Rogers, he was the first to describe the two cultures (Rogers 1939, 1945). The transition remains a debated topic to this day, and a number of hypotheses have been advanced to explain the period. Warren et al. (1961) and Warren and Pavesic (1963) proposed that the La Jollan culture began circa 7,500 years B.P. when desert gatherers moved west to avoid unfavorable Altithermal climates. Kowta (1969) outlined a similar hypothesis and claimed that the movement coincided with the diffusion of agave to the coast. Kowta strengthens these claims by suggesting that the primary function of the scraper assemblage was processing agave and yucca.

Moriarty (1966, 1967), Kaldenberg (1976), and Koerper et al. (1991), find that a continuity exists between the San Dieguito culture and the La Jolla culture. They claim that the La Jolla culture developed out of the earlier San Dieguito culture. Recently, Bull (1987) and Ezell (1987) have argued that the La Jolla culture and the San Dieguito culture are simply functional variants of the same culture. Others take the view that they are distinct cultures (Hayden 1987; Moriarty 1987; Smith 1987). Moratto (1984) suggests that a combination of some of these models can be used to explain the situation. "Climatic warming after circa 6,000 B.C. may have stimulated movements to the coast of desert peoples who then borrowed littoral adaptations from older groups while sharing with them their millingstone and scraper-plane technologies and seed- and agave-processing skills" (Moratto 1984:151).

Another hypothesis that has gained in popularity is coastal migration (Fladmark 1979). Chartkoff and Chartkoff (1984) claim that as coastal settlement is pushed further back in time, the more likely it is that the settlers were not related to the Pleistocene hunters of the interior deserts. Meighan (1989) has noted similarities between California lithic assemblages and assemblages in British Columbia and Alaska. Erlandson and Colten (1991), however, point out that the assemblages more closely resemble the interior desert assemblages. A major obstacle to solving this controversy is the fact that approximately 17,000 square kilometers of land has been inundated since the end of the last glacial (Carbone 1991), effectively burying most coastal sites. If Meighan's (1989) hypothesis is correct we would expect to find evidence of a coastal migration route at inundated offshore sites along the coast, and larger habitation sites with a material culture reflecting the transition period. This transition period is still a topic of debate and will clearly remain unresolved for quite some time.

Food Collecting Cultures (La Jolla, Pauma, and Sayles)

After a sketchy transition period, the La Jolla-Pauma cultures became established. The La Jollan culture is the project region's representative of Warren's (1967) "Encinitas Tradition" and Wallace's (1978) "Period II: Food Collecting." La Jollan sites tend to be located near the coast and particularly around lagoons and bays suggesting an ecological adaptation to shellfish and other coastal resources. Generally, the Encinitas Tradition in the San Diego area is dated from about 8,000 B.P. to 2,000 B.P. Pauma sites have a similar material culture, but lack shellfish and exhibit a more sedentary lifestyle (True 1958; Meighan 1954; Warren and True 1961). The Pauma culture has not yet been assigned firm dates (True 1980).

Rogers (1945) and Harding (1951) divided the La Jollan culture into two phases, La Jolla I and II. The latter phase was defined by the presence of cemeteries, trade with the Channel Islands, and an improved lithic technology. Later, Wallace (1955) assigned the culture to the "Milling Stone Horizon" and Moriarty (1966) divided the culture into three phases. Moriarty's La Jolla I (5,500-3,500 B.C.) is characterized by the appearance of milling stones, percussion-flaked scrapers, and flexed burials. La Jolla II sites (3,500-2,000 B.C.) are identified by the appearance of ground stone discoidals, cemeteries, and an increase in projectile point types. La Jolla III (2,000-1,000 B.C.) has a similar material culture but exhibits influence from the Yuman culture (Moriarty 1966:21-23).

The Sayles culture of the Transverse Ranges is thought to have begun sometime before 6,000 B.C. and continued until about A.D. 1,000 (Wallace 1955; Kowta 1969). Assemblages in this culture display a blending of the Pinto Basin culture of the Mojave Desert region and the Millingstone cultures of the coastal side of the Transverse Ranges (Moratto 1984). Late Prehistoric cultures of this area include the Serrano and the Cahuilla.

Diversified Subsistence (Late Prehistoric Cultures)

After an apparent hiatus (circa 1,000 B.C. to A.D. 500) in the cultural sequence or possibly a period of reduced activity (Meighan 1954; True 1958, 1966, 1970; Moratto 1984) there is an influx of Shoshonean (Takic) groups into the coastal zone (Moratto 1984). It is unclear if the hiatus reflects a migration out of the coastal areas or simply a period of reduced activity in these areas. The research design, outlined below (Chapter 4), takes this question into account and the archaeological investigations provide empirical data regarding this period.

The Late Prehistoric period in the southern California area is represented by two cultures in the coastal regions (San Luis Rey and Irvine) and one culture in the inland area (Cuyamaca). Originally defined by Meighan's (1954) survey data, the San Luis Rey (SLR) culture of San Diego County, was refined by True et al.'s (1974) excavation data, and has been equated with the historically known Luiseno (True 1966). While the original definition of the SLR culture was based on survey data, True et al.'s (1974) excavation data made only minor refinements in the definition of the culture (specific artifactual remains). Meighan (1954) and True et al. (1974) divide the culture into two phases, San Luis Rey I (A.D. 1400-1750) and San Luis Rey II (A.D. 1750-1850), based on the presence (SLR II) or absence (SLR I) of ceramics, cremations, and rock paintings. The Irvine culture of coastal Orange County is distinguished from the San Luis Rey culture on the basis of subsistence patterns and has been dated by Ross (1969) from A.D. 600-1800. The Cuyamaca culture of the southern San Diego County interior zones appears around A.D. 500 and can be traced to the historically known Diegueno (True 1970).

THE "CULTURE" CONCEPT AS USED IN CALIFORNIA ARCHAEOLOGY

With the cultural chronology loosely established, a critical examination of the factors used to draw cultural boundaries is now provided. In southern California, the determination of cultural boundaries is generally based on the issues of settlement, subsistence, and other economic factors (such as trade and technology). According to the proponents of the cultural sequences advanced for California (e.g., Rogers 1945; Wallace 1955, 1978; Warren 1968), subsistence and technology changed in response to individual cultural adaptations to the changing environment. Thus, early cultures relied primarily on game animals until environmental changes reduced the availability of these resources. Later cultures gathered plants and/or used coastal resources depending on their location. Finally, the Late Prehistoric cultures exploited a variety of available resources. As discussed above, however, cultural boundaries are generally based on subtle differences. This section critically examines the changes in settlement and subsistence patterns used by archaeologists to construct "cultural boundaries" in southern California.

Beyond simply discussing the paleoenvironmental conditions, a brief explanation of how cultures adapt to their environment is necessary. These changes in adaptive strategies have been modeled by a number of archaeologists who subscribe to a cultural ecological view of change (Jochim 1976; Yellen 1977; Binford 1978, 1980, 1982; Bettinger 1979, 1980; Thomas 1972, 1973). Most of these studies are based on data gathered from the Great Basin and are variants of Steward's (1938, 1955) cultural ecological model of culture change. Essentially, Steward argues that the unpredictable environmental conditions of the Great Basin combined with the simple technology of the cultures leads to very flexible subsistence patterns. The particular resource exploited by a group, at any given time, depends on that resource's availability at the time. This type of subsistence leads to a settlement pattern consisting of small temporary camps with more permanent camps inhabited at times when the individual groups meet for the purpose of social and economic exchange. Changes in environmental factors lead to an almost constant adjustment on the part of hunter-gatherer cultures, and minor changes in adaptive strategies take place on an almost daily basis.

If the cultural ecological model of adaptation is followed, distinct culture changes would not, by definition, exist. Thus, employing the cultural ecological model to reinforce the proposed cultural sequences in southern California is a poor tactic. We follow the cultural ecological model of change, and call for some revisions in the proposed cultural sequence (see Chapter 11). The individual adaptive patterns of subsistence for each of the cultures in southern California, as described by the proponents of the existing sequences, are now examined.

Hunting Cultures

The transition from the Pleistocene to the Holocene (12,000 to 8,000 B.P.) was a time of major environmental changes. Although varying in magnitude and duration, warming trends in the Holocene led to the evaporation of pluvial lakes, rising seas, changes in drainage patterns, and changes in both the flora and fauna of southern California (Antevs 1953; Deevy and Flint 1957; Van Devender and Spaulding 1979; Axelrod 1981; Koerper et al. 1986; Glassow et al. 1988; Carbone 1991). This changing environment affected the human population and led to a number of adaptive responses that often appear in the archaeological record as visible cultural differences.

Inland San Dieguito sites tend to be located on terraces above ancient lake beds and marshes, while coastal sites are generally located in and around estuaries and river channels. These locations appear to reflect an early adaptation to the abundant and easily exploitable resources of these settings (Warren and Pavesic 1963; Weide 1968; Moratto 1984; Warren 1987). Hayden (1987) claims that the San Dieguito culture covers an area extending from the California desert region east to Texas, and as far south as Oaxaca, Mexico. As discussed above, this culture is divided into three phases, based on technological differences and site location. The lithic assemblage is more varied and sophisticated in the later phases. Percussion- and pressure-flaked tools are common and include points, knives, crescents, scrapers, choppers, and hammerstones. Ground stone artifacts are noticeably absent from the early assemblages but start to appear on later sites, reflecting adaptive responses to the changing environment.

Presenting a cultural ecological view to explain these later changes, Koerper et al. (1991) claim that the presence of milling equipment and shellfish remains on a number of sites has changed the view of the San Dieguito culture from a generalized hunting tradition to one of more diverse subsistence strategies. More recently, Jones (1991) claimed that the archaeological record shows great variation in the use of marine resources. Marine resource use varies with a number of critical variables including coastal environmental conditions and the efficiency ranking of the locally available coastal resources compared to terrestrial resources. Essentially, Jones challenges established beliefs surrounding the subsistence patterns of the first inhabitants of California, claiming that coastal resources were important in some regions. If these views are followed, identifying a distinct transition from San Dieguito to La Jolla would be an extremely difficult task. In addition, if the major transition period is difficult to identify, the proposed, subtle differences between phases within these cultures would be nearly impossible to spot. Again, the paucity of investigated, transitional sites has impaired the archaeologist's view.

Food Collecting Cultures

Proponents of existing cultural sequences claim that subsistence patterns show marked changes starting around 6,500 B.C. Whether these changes are the result of the adaptation of cultures already inhabiting the region or due to an influx of peoples from the east, or a combination of both, is still a controversial topic. However the changes occurred, they are almost certainly in response to Altithermal climatic conditions and the changing flora and fauna (see Gallegos and Hector 1987; Erlandson and Colten 1991). These changes are visible in the archaeological record as a reduced number of projectile points, scrapers, and choppers, and an increase in the amount of ground stone artifacts. While hunting and fishing were not replaced by hard seed processing, the reliance on animals and fish decreased and the diet became increasingly diversified (Koerper 1981). Diversity of adaptation appears to be the norm of the Encinitas tradition (Millingstone horizon). Within this extremely diverse tradition, however, archaeologists have identified a number of "distinct" cultures.

La Jolla Culture

The La Jolla culture exploited the coastal regions of Orange and San Diego County. La Jollan sites are recognized by ground stone assemblages in shell middens, usually on terraces around lagoons or bays (Rogers 1939, 1945; Harding 1951; Moriarty 1969; Quillen et al. 1984). Major La Jollan sites identified in the San Diego area include Fairbanks Ranch (W-595, W-1353), Batiquitos Lagoon (SDI-603), Santa Fe Knolls, Scripps Estates, and SDI-5130, all of which are on or near a lagoon.

The village site SDI-5130, is located approximately one mile downstream from the project area. Data recovered from this site suggest a period of occupation from 5,000 to 6,000 years B.P. and display a subsistence pattern based primarily on shellfish and plant foods supplemented with small mammals (Quillen et al. 1984). Trade beads and obsidian sourcing data indicate that the village had trade relations with groups to the north, south and east. Quillen et al. (1984:5) propose that the abandonment of the site was an adaptive response to the sedimentation of the San Luis Rey Lagoon.

The technology at La Jollan sites indicates a mixture of coastal and desert traits since both scraper-plane and ground stone artifacts are found. La Jollan sites include shell middens, fire hearths, ground stone, flexed burials, and a very basic lithic assemblage. In addition, the tool assemblage appears to indicate a greater reliance on marine resources than the San Dieguito culture (Kaldenberg and Ezell 1974). As indicated earlier, however, Kowta (1969) has suggested that the scrapers were most likely used to process agave and yucca (see also Salls 1983; Basgall and True 1985). In addition, it has been suggested that these tools were used as: woodworking tools (Eberhart and Wasson 1975), hammers (Johnson 1966), resharpening tools (Treganza and Bierman 1958), flake tools (Jackson 1977). Some of the more interesting artifacts representative of this culture are discoidals and cogged stones (often found together). Although many uses have been proposed, the actual function of these stones is unknown (Eberhart 1961; Dixon 1968, 1975; Moriarty and Broms 1971). In the end, it appears that La Jollan assemblages represent a transitional phase from San Dieguito to Late Prehistoric cultures rather than a culture with distinct spatial and temporal boundaries. Additional evidence that indicates the unclear spatial boundaries of the La Jollan culture is the presence of the very similar Pauma culture, discussed below.

Pauma Culture

The Pauma culture is similar to the La Jollan culture in many ways. The main difference between the two cultures is that Pauma sites are generally found in inland valleys and sheltered canyons, out of reach of marine resources. True (1958), Warren et al. (1961), and Meighan (1954) describe Pauma sites as those which display a relatively more sedentary lifestyle and a greater reliance on gathering when compared to the San Dieguito culture. Pauma sites also contain many ground stone artifacts, a great tool variety, and lack shellfish remains. Artifact assemblages are similar to La Jollan sites, but subsistence practices are apparently more focused on terrestrial resources. D.L. True described what is know about the Pauma culture in the following way:

> (1) The Pauma Complex inventory is very similar to that of the adjacent coastal La Jollan...and some undefined but close relationship is proposed between the two; (2) the Pauma Complex as defined here includes very little actual evidence of San Dieguito elements...; and (3) there may be evidence in the area (sometimes associated with Pauma Complex sites) of the Campbell intrusion proposed by Warren (True 1980:30).

Even with these recent clarifications by True (1980), many inland, non-ceramic sites continue to be automatically classified as Pauma (Carrico 1987).

The difference between La Jollan and Pauma sites is clearly based on functional differences in the artifact assemblages. A greater tool variety in Pauma sites indicates a greater reliance on terrestrial resources than La Jollan sites. This adaptation is most likely the result of terrestrial resource availability in the settlement area rather than cultural differences. It appears that any "close relationship" between the two could be explained by viewing the two cultures as functional variants of the same culture. The classification of these two groups as separate "cultures" seems to be the result of archaeologists attempting to define distinct cultural boundaries in an area where boundaries are very blurred.

Sayles Culture

Between the Mojave Desert region, exploited by the Pinto Basin culture, and the coastal area, exploited by the Millingstone cultures of southern California, is a culture that exhibits an assemblage of artifacts that appears to represent a utilization of resources from both the desert and coastal regions. Archaeological investigations, beginning in the 1940s in the Cajon Pass region of southern California, revealed the type site (SBR-421) for the Sayles culture and a number of other sites that typified the culture (Moseley and Smith 1962; Bowers 1976). Excavations at SBR-421 were conducted by Kowta (1969) in 1965 and 1966. The excavations recovered a millingstone assemblage that also included percussion-flaked scraper planes, cores, plano-convex scrapers, choppers, and hammerstones. The mixing of materials within the assemblage indicated to Kowta (1969) that sites of this culture represent a period of technological change. For Moratto (1984), this culture shows evidence of blending between the Pinto Basin culture of the Mojave Desert and the Millingstone cultures of southern California.

Subsistence patterns, as indicated by archaeological investigations, are based on opportunistic hunting of deer, rabbit, and other small game animals, as well as floral resources such as juniper berries and hard seeds (Basgall and True 1985). Whereas investigations suggest that resources were available on a year round basis, a limited variety of tools within artifact assemblages suggests a more seasonal use of sites. From the limited information known about the Sayles culture, it seems likely that the culture is a representation of the transitional period between the early hunting cultures and the later gathering cultures. Again, any temporal distinction between the Sayles culture and other cultures is most likely an arbitrary midpoint where differences are most apparent. From this discussion it should also be apparent that any spatial boundaries drawn between these cultures are most likely hypothetical lines proposed by archaeologists.

Diversified Subsistence (Late Prehistoric Cultures)

Late Prehistoric cultures in southern California most likely reflect both in situ cultural adaptations in response to environmental changes and outside influence from the Shoshonean (Takic) intruders of the desert regions (Moratto 1984). In southern California, three "distinct" cultures are commonly identified: The San Luis Rey culture of San Diego County, the Irvine culture of Orange County, and the Cuyamaca culture of the inland desert region of San Diego County. As is true in the earlier periods, cultural distinctions are often blurry and/or based on subtle differences. Below, each culture is briefly discussed.

The San Luis Rey Culture

As discussed above, the San Luis Rey culture of San Diego County was originally defined by Meighan (1954) at SDI-132, and later refined by True et al. (1974) at Molpa, (SDI-308). Based on the results of numerous surveys (Eberhart 1952; Meighan 1959a; Warren and True 1961; Warren et al. 1961), and excavations, Meighan (1954) and True et al. (1974) divide the culture into two phases, San Luis Rey I (A.D. 1400-1750) and San Luis Rey II (A.D. 1750-1850), based on the presence (SLR II) or absence (SLR I) of ceramics, cremations, and rock paintings.

The type site for the SLR I phase is SDI-132. At this site, 35 km from the coast, bedrock mortars and associated millingstones were found. The assemblage also included triangular projectile points, bone awls, and assorted stone and shell artifacts. In addition to the features and artifacts, a number of cremations were also discovered (Meighan 1954). The type site for the SLR II phase is Molpa (SDI-308). While subsistence patterns are relatively constant over the two phases, SLR II adds pottery vessels (including cremation urns), pictographs, and nonaboriginal items such as glass beads and metal knives. A relatively sharp distinction can be drawn between SLR I and II, based on material culture. Whether these distinctions reflect important cultural changes or simply temporal differences remains to be seen.

Based on the information gathered on the above-mentioned surveys and the excavations at SDI-132, it was inferred that the culture's subsistence pattern was small game hunting and the gathering of seeds and nuts, especially acorns (Meighan 1954). As contact between the aboriginal cultures, missionaries, and other settlers increased, the subsistence patterns of the San Luis Rey culture began to incorporate non-native plants and animals and focus less on coastal resources (Kroeber 1925; Strong 1929; Bean and Shipek 1978; Moratto 1984). Based on ethnographic and ethnohistoric accounts of early contacts with the culture, the settlement pattern was similar to the later Luiseno rancherias. Small settlements were located at spots from the river basin to the higher mountain slopes and were occupied on a seasonal basis depending on resource availability. It is relatively certain, based on archaeological and ethnographic evidence that the San Luis Rey culture depended on a variety of resources ranging from sea mammals to acorns.

The Irvine Culture

The Irvine culture is very similar to its neighbor, the San Luis Rey culture. According to Ross (1969) the main difference appears to be economic. Irvine culture sites tend to reflect a high degree of dependence on marine resources. A shell midden site in the Newport Bay area (ORA-190) revealed 65 species of marine invertebrates, three marine mammals, and 11 different species of fish (Ross 1970). Occupation of ORA-190 dates to 1,350 B.P.

Settlement patterns are also similar to the San Luis Rey culture in that small rancherias are found in river basins and on mountain slopes. The distinction between the two cultures appears to be similar to other distinctions drawn in the southern California area; boundaries are arbitrarily drawn and reflect spatial and/or functional rather than cultural differences. Major excavations along the Irvine coast have been underway since 1987, but published reports are not yet available. It is anticipated that these excavations (including a major village site and numerous smaller habitation sites) will help to further define both the La Jollan and Irvine cultures (R. Mason and H. Koerper, personal communication 1990).

The Cuyamaca Culture

The Cuyamaca culture inhabited the inland desert regions of southern San Diego County. True (1970) defined the culture as distinct from the San Luis Rey culture and placed it temporally between the early La Jolla and modern Diegueno cultures. True (1966, 1970) distinguished this culture from the San Luis Rey culture based on cremations that were placed in urns and buried in cemeteries away from habitation areas. Also, the Cuyamaca culture had a steatite industry and used a wider variety of ceramics than the San Luis Rey culture (Moratto 1984:156). "Urn burial of cremations in defined cemeteries, manufacture and use of specially made items for mortuary use, such as miniatures (pipes, ceramic vessels, and shaft straighteners), possibly specially made projectile points, and the idea of the clay lined hearth, as well as an emphasis on manufacture and use of ceramics, as compared to adjacent groups (Luiseno and Cahuilla) all seem to indicate Colorado River ties" (True 1970:59). True (1970) indicates that subsistence patterns of the Cuyamaca involved the use of winter and summer villages where the exploitation of acorns and other vegetable resources was supplemented by small game animals, insects, and eggs.

True (1970) distinguishes the Cuyamaca culture from the San Luis Rey based on stylistic differences rather than functional artifact classes. This differentiation establishes a boundary that appears more cultural than most of the other functional distinctions in the southern California region. It appears likely that this group migrated from, or was heavily influenced by, the Colorado River cultures. The Cuyamaca culture has many similarities to the historically known Diegueno culture, which is clearly a separate culture from the Luiseno.

Integrating Settlement, Subsistence, and Chronological Issues in Prehistory

Artifact assemblages throughout the prehistory of the San Diego region show little change over time. Archaeologists have generally interpreted this fact as representing long periods of cultural stability. When change is evident in the archaeological record, this is generally interpreted as evidence of dramatic culture change or interaction with outside groups. For example, one generally accepted view of the onset of the Encinitas Tradition (Millingstone Horizon) is that the drastically changing environment dried up the pluvial lakes, changed drainage patterns, and led to a reduction of inland resources. The human adaptive response to the environmental changes was to move to the coast where resources were abundant. Likewise, cultural change in the San Luis Rey River basin and surrounding region has been connected with the sedimentation of lagoons over the past 2,000 to 3,000 years (Warren et al. 1961; Warren 1964; Gallegos 1985, 1987). One of the major goals of this project, related to this topic, was to determine if the site was located on the hillside above a lagoon or a river, in order to more accurately reconstruct the adaptive patterns of the inhabitants. The results, discussed in Chapter 11, have added new information to issues regarding cultural change and continuity in the San Diego region.

While it is true that the limited view of the archaeological record seems to display distinct cultural boundaries between assemblages, it also holds that long periods of stability, punctuated by periods of rapid change, goes against most models of culture change advanced for hunter-gatherer groups. Change in groups at this level of social organization is usually the result of a changing economic emphasis. Innovation and invention are far less likely explanations. Thus, change in the San Diego region was most likely slow and gradual, taking hundreds or even thousands of years to unfold. As a result of slow and gradual change, most of the temporal and spatial cultural boundaries, drawn by archaeologists in recent years, can be viewed as simply arbitrary divisions along a continuum. In addition to arbitrarily drawn cultural boundaries, many post-depositional formation processes have mixed cultural layers and destroyed distinct stratigraphic horizons to the degree that, even if clearly distinct cultures existed, they would be difficult to identify because they are often found at the same depth (see Schiffer 1976, 1983, 1987, for a complete discussion of formation processes). As discussed by Altschul et al. (1984:109), changes must be viewed as a "dynamic series of events." Detecting the slow patterns of change or stability over long periods of time has been a major goal of much archaeological research and is one goal of this project.

Based on the discussion above, use of the "culture" concept in California appears to have obscured the archaeologist's view of culture change and continuity. Compartmentalizing prehistoric populations into distinct cultures based on functional differences in technology and other environmental adaptations serves little purpose beyond establishing one archaeologist's view over another's. It is our opinion that while the "culture" concept serves as an important classificatory device in many parts of the world, it is of little utility in southern California. Unfortunately, we are left with a number of vague cultural sequences proposed during the past fifty years, and must acknowledge these traditional schemes (see Chapter 11 for our views on culture change).

HISTORY OF THE SAN LUIS REY REGION

Ethnohistoric and Ethnographic Background

While prehistoric cultural differentiation has been shown to be an extremely tenuous task, dividing historic cultures would appear at first glance to be much simpler. Assigning a contact date, when the Spanish first encountered the Native American population is easy. Identifying internal cultural change after contact is, however, more difficult. Some of the early encounters led to acculturative processes, whereas others had little impact on the indigenous population. Thus, the transition from the prehistoric to the historic period is simple to date, yet difficult to interpret.

The earliest historic account of contact with the Luiseno was in 1540, when Francisco de Ulloa, a lieutenant under Cortez, went ashore at the mouth of the San Luis Rey to fill his canteens and ask the Indians if they knew of any overland routes to the seven cities of Cibola (San Luis Rey Historical Society 1957). His ability to communicate was limited, so he continued north along the coast. Finding no other promising route, he returned to San Luis Rey about one month later. The following account is given:

> His second disembarkment included his entire ship's company with arms, food, jewels and gold and silver booty from Mexico. It was accomplished on August 31, 1540.
>
> For the first week they camped near the Indian villages that dotted the shore like a string of beads.
>
> From the filth of the lake and villages the Spaniards lacking the immunity built up over the centuries by the Indians, acquired and rapidly succumbed to a prostrating and fatal form of dysentery.
>
> Captain Ulloa aware, too late, that the disease afflicting his men was associated with the villages at the end of 10 days moved away from the lake and into the protection of a primitive fortress and pre-historic battle-ground. The giant boulders of this area grouped by nature to protect ancient warriors from hostile invaders, bear the symbols and inscriptions of men who lived, according to authorities, as early as 50,000 years ago.
>
> After 3 days illness in this ancient fortress, Ulloa died and was buried by his men on September 5, 1540. The site is marked with a white cross (San Luis Rey Historical Society 1957).

The "ancient fortress" site (SDI-8729) described in the above account is located about two miles northeast of SDI-6010. Although the site was investigated in the 1950s, the grave of Ulloa was never discovered.

Other early historic accounts of contact with the Luiseno are in 1542 with Cabrillo and in 1602 with Vizcaino. Notable ethnographic accounts are by Henshaw (1972), Sparkman (1905, 1908a, 1908b), Du Bois (1904, 1908), and Kroeber (1906, 1908, 1909, 1917, 1925). Later, numerous scholars (Gifford 1918, 1922; Strong 1929; Harrington 1933, 1934; White 1963) added to the ethnographic literature.

While European exploration of California began in 1542, with the arrival of Juan Rodriguez Cabrillo, the Spanish presence was not felt in San Diego until Don Gaspar de Portola's overland expedition encountered the Luiseno in 1769. The Spanish found the San Luis Rey River valley to have a plentiful water supply and lush vegetation (Bolton 1926). According to Bolton (1926), the explorers also found that two large Luiseno villages were situated on both sides of El Camino Real at each end of the San Luis Rey River valley. It was not until 1776 that the Mission San Luis Rey was founded and serious cultural impacts began.

According to Bean and Shipek (1978), Luiseno territory extended approximately from Agua Hediona Creek northwest to Aliso Creek along the coast, then east to Santiago Peak and south through the Lake Elsinore area to just south of Mount Palomar (Figure 8). "To the northwest and north they had Juaneno, Gabrielino, and Serrano as neighbors; to the east the Cahuilla, and to the south the alien Diegueno of Yuman family" (Kroeber 1925:648). The Luiseno, through complex social organization and mechanisms such as clan-governed districts and seasonal movement of populations throughout the region, followed a planned program of resource utilization to exploit the abundant plants and animals and support a large population (White 1963; Kroeber 1925; Strong 1929; Bean and Shipek 1978). The seasonal exploitation of acorns and small game (on the inland hill slopes) was combined with the exploitation of coastal resources during the balance of the year. Subsistence also included coastal marine resources and fish from the rivers and streams (Sparkman 1908). According to Sparkman (1908), fish were the main resource of the coastal inhabitants:

They used a canoe or raft of rushes, with which they went out some distance from the shore to fish with a dip net. Seine nets were also used. Some wooden canoes were also made from the trunks of trees. It is stated that voyages were formerly made with these as far as San Clemente Island. The coast people also fished with the hook and line. The line was made from the fibre of *Yucca mohavensis* and the hook from abalone shell... Fish hooks were also made of bone. The Coast people also consumed large quantities of shellfish of several species. Some say that they used a harpoon for spearing fish... Others say that no harpoon was ever used by them. This may be true, but it is certain that the Diegueno used one, and it seems improbable that the Luiseno would not have employed it (Sparkman 1908:200).

Luiseno villages were usually located in defendable canyons or coves along the slopes near good water supplies (Bean and Shipek 1978). Population for the villages ranged between 50 and 200, with the larger ones spawning nearby satellite villages. Kroeber (1925) estimated the ancient Luiseno population to be around 4,000 and the 1925 population to be less than 500. In contrast to Kroeber's estimation, White (1963) placed the population at approximately 10,000 (based on 50 villages with an average of 200 people each). Finally, Cook (1976) claims that the aboriginal population of California stood at 310,000. Based on her estimate of three people per habitable square mile, the Luiseno population would stand at about 4,500. It is likely that the true population of the early cultures will never be known, but it is probably a safe approximation to place the population somewhere between 4,000 and 10,000. However large the ancient population was, it rapidly decreased after contact due to the spread of disease and a sharp decline in living conditions (Bean and Shipek 1978).

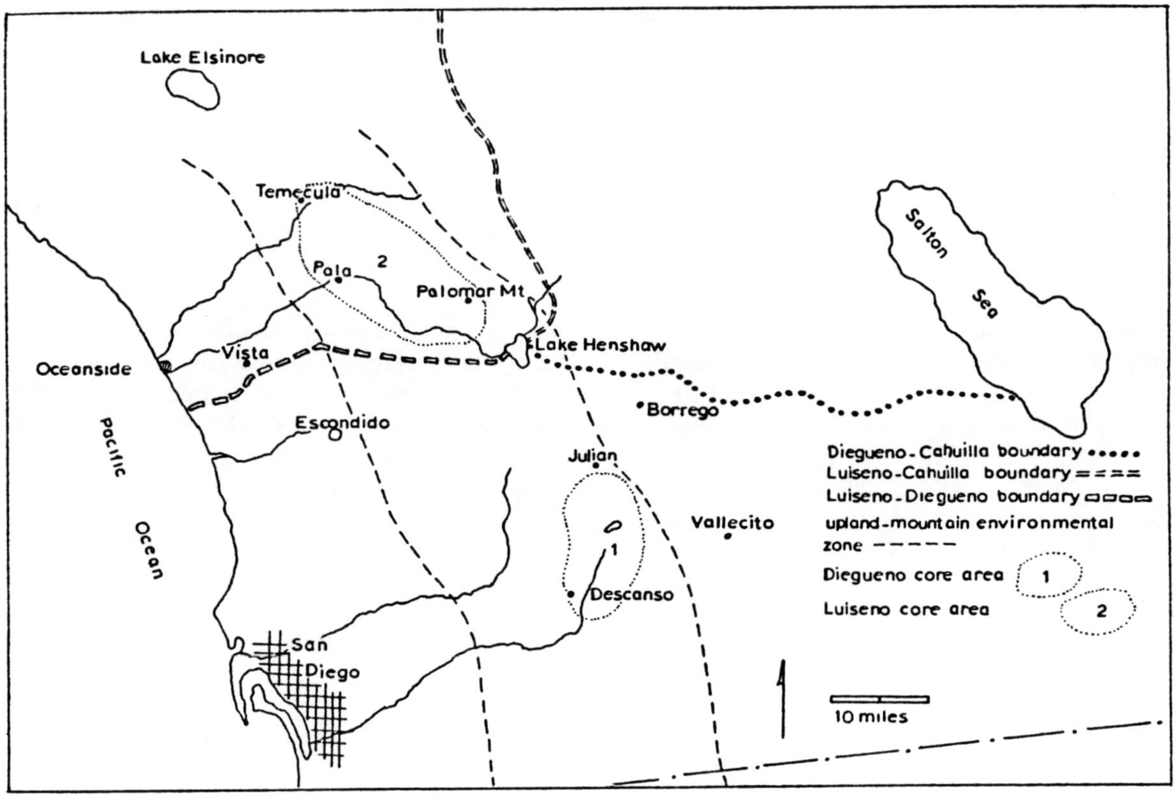

Figure 8. Map of Luiseno territory and surrounding area (from True 1966).

Mission San Luis Rey

The Spanish, interested in establishing a mission between San Diego and San Juan Capistrano, were drawn to the San Luis Rey River valley because of the water supply, abundant vegetation, and the large native population (Engelhardt 1921). Mission San Luis Rey was established in 1798 and quickly became one of the most productive missions in California (Bolton 1926). Many Native Americans were brought to the mission, where they were taught the faith, the Spanish language, and crafts (Bean and Shipek 1978). For the most part, however, the Luiseno maintained their previous settlement patterns and political leadership. The success of the mission began to decline in 1833 when a decree of emancipation of the Indians was passed and in 1835 the mission was confiscated by the Mexican Government. At this time land was granted to citizens for use as grazing land (Elliot 1883; Moyer 1969).

The Rancho Period

A number of factors led to the deterioration of Native American lifeways. The Gold Rush and the granting of statehood to California brought many Anglo-Americans to the area (Bancroft 1886; Kroeber 1925). Additional stress came with the secularization of the mission and the lands being split up; the local Indians were forced to either work on ranchos or become rebels (Moratto and Greenwood 1991; Gibson et al. 1990). Rancho Guajome, 2.5 miles east of the mission was given to Cave Couts in 1851 who then, using Indian indentured servants as laborers, soon became one of the most important figures in the area. Later, Couts entered into a business agreement with Don Juan Forster, the owner of Santa Margarita and Las Flores Rancho, and together these two ran cattle on public lands in the area (San Diego County 1854, 1867). After the Civil War, the influx of settlers to the area began to erode the power of the ranchers. The farmers had also gained power, and the fence laws, passed against the ranchers, increased the costs of cattle ranches (Caughey 1953; Moratto and Greenwood 1991).

Additional important settlers in the area include: Herbert Crouch, a sheep rancher; Simon Goldbaum, who established a farm in the valley in 1871, and owned some businesses in San Luis Rey (Elliot 1883); and Matthew Hubbert, who moved his family to the valley in 1873 and settled on land purchased from the Fletcher family (Hubbert 1936; Crouch 1915; Kyle et al. 1988). According to Franklin and Carrico's (1979) investigations of the historical record, when Hubbert moved to this property in 1873, a Native American rancheria was located there. Hubbert reportedly tore down the structures and reused the tiles and bricks to construct his own structure. The Luisenos reportedly moved to an area near Whelan Lake (believed by Franklin and Carrico to be SDI-6009). While a number of adobe structures were built on the ranchos and farms in the valley, very little evidence remains at the locations indicated from maps and bibliographies. The present site, SDI-6010, had a corral located on it, the details of which are investigated in Schmidt et al. (1992).

CHAPTER 4
RESEARCH DESIGN

INTRODUCTION

The purpose of a research design is to outline a set of goals that we will attempt to achieve through implementation of a specific method. The goals must be specific enough to guide the investigators through the project, yet flexible enough to allow for unexpected finds and other difficulties. A research design is essentially a tool used by the profession to ensure that archaeology is not simply a way to collect artifacts, but rather a scientific process carried out to gain knowledge about the relationship between material culture and past human behavior (Schiffer 1976, 1987) and, in the end, about the present human condition (Hodder 1985, 1986; Leone 1981; Shanks and Tilley 1987).

The goals of this testing phase were established to assess the site's eligibility for inclusion in the National Register of Historic Places. This chapter outlines the research design employed by SRI at the prehistoric component of SDI-6010. The research design guiding the historical archaeological studies was set forth in Schmidt et al. (1992). Our research design is organized into three main sections. First, a brief discussion of the defining characteristics of a research design and the scientific method in archaeology is provided to familiarize the reader with the process of constructing meaningful goals. Second, the specific goals outlined for this project are explained with reference to the existing database of archaeological and environmental knowledge discussed in previous chapters. Test implications, critical for hypothesis testing, are spelled out in detail after each set of questions. Next, the methods, procedures, and techniques used during the investigations are explained. The second section also explains the inherent biases associated with each particular method, and provides the reasons behind particular methodological choices. Finally, the chapter concludes with a critical review of this research design to explore some of the biases inherent in our interpretations. If these biases can be recognized before the investigations begin, we can attempt to avoid certain pitfalls in our interpretations.

THE SCIENTIFIC METHOD IN ARCHAEOLOGY

While the ultimate goal of archaeology as a scientific endeavor is debatable, many archaeologists agree that, at a minimum, it seeks to illuminate "the conditions that brought about change and modification in the organization of dynamics occurring in past living systems" (Binford 1977:7). Most archaeologists are not interested in the artifacts themselves, but rather the society that crafted, used, and left behind those artifacts. The only way to gain knowledge about past societies is through the process of inference. "An inference is a statement about the past supported by relevant principles and relevant evidence" (Schiffer 1987:4). Research designs serve to guide archaeologists through this process of inference toward the goals of their particular investigations.

Rathje and Schiffer (1982) outlined eight "research stages" to the scientific method employed by archaeologists to gain insights into the relationship between material culture and human behavior. The process is constructed on a "conceptual scheme," or the major assumptions upon which the discipline defines its direction of research and broad areas of interest. First, drawing on previous investigations, the researcher formulates specific research questions. Next, the researcher provides multiple answers to the questions, phrased in the form of hypotheses. If a particular hypothesis is true, then it should have a set of expected outcomes. "*Test implications* state these expectations and clearly specify the *relevant data* - those specific observations that are crucial for hypothesis testing - that need to be collected" (Rathje and Schiffer 1982:40, emphasis original). After the research questions, hypotheses, and test implications have been established, the investigator then proceeds to the data recovery phase (survey, excavation, experiments, etc.). After the data have been collected, they are analyzed and then synthesized and assessed. The final stage is determining the results and, if necessary, rethinking the hypotheses. It is this process that allows the archaeologist to work with two methods of discovery of data and hypotheses, "inductivism" and "deductivism." Binford (1972), Martin (1971), Dunnell (1982), and Watson et al. (1971) claim that archaeologists must work back and forth between induction and deduction, in order to generate new hypotheses and confirm existing ones (Kelley and Hanen 1988).

We follow the scientific method outlined by Rathje and Schiffer (1982) by formulating hypotheses based on prior research conducted in the area. After testing these hypotheses through the excavation, analysis, and synthesis of data, we will evaluate the site's significance and, if necessary, create a research design for data recovery based on these hypotheses or, if necessary, new ones. Thus, we are working toward the creation of a research design for future investigations from two sides. First, by reviewing past investigations in the area, we formulate an "educated guess" (hypothesis) based on what we expect to find. Second, after we test the site, we use the data to alter our hypotheses. In this way, induction and deduction work together toward a common goal.

Broad, sweeping research questions, addressing issues such as culture history, paleoenvironmental conditions, site context, and settlement patterns are outlined below, but since the present investigation is simply a test of the site's integrity and ability to address these important issues in prehistory, more specific research questions are necessarily included. During this testing phase of the archaeological investigations, all eight stages of research are followed in order to assess the site's eligibility for inclusion in the National Register, and if the site is deemed eligible, to construct a treatment plan and research design for future investigations. If the site is, in fact, deemed eligible, the broad research questions will be addressed in later investigations.

In Chapter 2, we discussed the environmental background of the project region. Chapter 3 outlined the culture history and ethnographic framework as well as the history of previous archaeological research in the San Luis Rey region. This information provides the background for the formulation of a series of research questions designed to alleviate gaps in our present understanding. Research questions pertinent to the prehistoric and historic components of SDI-6010 are presented separately. These questions are discussed in the next section.

RESEARCH QUESTIONS AND DISCUSSIONS

General Questions

In 1989, the COE (Los Angeles District), the Advisory Council on Historic Preservation (ACHP), and the California State Historic Preservation Officer (SHPO) entered into a Programmatic Agreement regarding the treatment of cultural resources that may be affected by the San Luis Rey Flood Control Project. Taking into account the stipulations outlined in the Programmatic Agreement, previous research, current understanding of the prehistory and history of the region, environmental background, and the scope of work for this project (Corps Of Engineers 1991), three basic research goals are defined:

(1) To assess the eligibility of SDI-6010 for the National Register of Historic Places (cf. 36 CFR 60.4);

(2) To determine whether human remains are present; and

(3) To gather information necessary to formulate a treatment plan and research design to guide future data recovery mitigation.

The most basic goal of the testing program at SDI-6010 is to determine whether the site is eligible for listing in the National Register of Historic Places. The testing program also needs to be of sufficient detail that, if the site is considered eligible for the National Register, a treatment plan can be devised. For National Register eligibility, the evaluation of SDI-6010 will most likely be determined under Criterion D of 36 CFR 60.4. Although it is possible that the historic component could qualify under Criteria A or B, given current knowledge, this is highly unlikely.

The issue of eligibility to the National Register in terms of Criterion D is relatively straightforward. It requires an assessment of scientific importance or "significance" of the cultural resources concerned (see Butler 1987). Determining what is significant is less easy, however. There is no formula or "cookbook" approach to making this assessment, nor should one exist. National Register eligibility is "specific to the area in question and the theoretical and substantive knowledge of the discipline [archaeology or history]" (Butler 1987:821). Thus, decisions about a property's significance can reliably be made only within the context of regional history (National Park Service 1986:9); compiling background history or "historical context" is the first logical step in designing any research, and is crucial to the evaluation of any property. All eligibility determinations must state *why* a site is or is not important in relation to current regional research questions, and provide adequate data to support the subsequent recommendations of eligibility.

The approaches to evaluation of the historic and prehistoric components of SDI-6010 are parallel. To evaluate its National Register status, we must first develop regional research questions that we might expect to address with data recovered from the site. These questions are grounded in the prehistory and history of the area and in previous scientific investigations. Once these questions are posed, we must then develop a research strategy that will (1) identify the types of data needed to answer these questions; (2) determine the most efficient data acquisition strategy; (3) determine whether the necessary data are found at the site; and (4) analyze the adequacy, integrity, context, and relevance to the research questions of these data. Finally, based on the analysis of the data obtained during the testing program, we will determine whether the site has the potential to yield *important* information -- that is, can it address the research questions posed. If the answer is "yes", then the archaeological site will be recommended to be eligible for listing in the National Register of Historic Places.

It is worth noting that, whereas an evaluation of eligibility is made for the site as a whole, each individual component may or may not be significant. For example, we may find that while the historic deposits at SDI-6010 are well-preserved and possess the potential for providing crucial information on the history of the San Luis Rey region, the prehistoric component has been disturbed by subsequent historic occupation, especially farming, and little intact subsurface deposits remain. In this situation, although the site would be determined eligible for inclusion in the National Register, only its historic component would be considered significant. Thus, whereas determinations are made about sites as a whole, research strategies must be designed to evaluate each component individually.

The Whelan Lake site contains both a prehistoric and historic component. The two are culturally distinct, with their only link being spatial proximity. Because they are distinct, we decided to treat them as separate entities, with separate evaluations conducted for each. The discussion herein deals only with the prehistoric component at SDI-6010. The reader is referred to Schmidt et al. (1992) for an evaluation of the historic component.

General Methodology

To obtain the requisite data, a comprehensive search of the entire parcel needed to be undertaken. While the entire site was surface collected, modern disturbances made the surface observations inadequate to properly delineate site size and identify previously unrecorded resources. Instead, a form of subsurface probing had to be devised. In essence, probing became a form of three-dimensional survey. Much like a pedestrian survey, our objective was to inspect the parcel using transects spaced at specified intervals. Instead of walking each transect, however, we used a backhoe to excavate a trench, thereby providing a linear exposure.

Our strategy was to place transects at specified intervals over the entire property. Whereas trenching provided both the vertical and horizontal parameters of the various prehistoric and historic deposits, this strategy did not supply sufficient information on site content or integrity. To obtain data on these issues, 10 1 m x 1 m and two 2 m x 2 m test units were manually excavated in the identified cultural deposit. Other samples that were taken to answer questions concerning chronology, subsistence, paleoenvironmental issues, and other areas of interest include pollen, radiocarbon, flotation, and obsidian hydration samples. All materials were processed and catalogued. Materials from the project are curated at the San Bernardino County Museum.

Specific Prehistoric Research Domains

Paleoenvironmental Issues

Since few well-documented paleoenvironmental reconstructions exist for the lower San Luis Rey River basin (Warren and Pavesic 1963; Gallegos 1987), one of the most important goals for this project is to establish the environmental conditions of the site during its occupation. At approximately 7,500 to 8,000 B.P., coniferous vegetation had been almost completely replaced by oak woodland and sage scrub. Coastal archaeological sites from about 7,000 B.P. show *Chione* sp. beginning to outnumber *Argopecten* sp., possibly indicating a shallowing of lagoon environments as the sea level began to rise (Gallegos 1987). Pollen and sedimentation studies of coastal marshes are critical for an accurate reconstruction of the paleoenvironment (see Mudie 1976; Koerper 1981). Such studies by nature involve large numbers of variables that are difficult to control. Even so, a tentative reconstruction provides a heuristic model from which future archaeological studies can build. Specific questions to be investigated are outlined below.

Research Questions

1. Was the site located above a lagoon or a river? What was the local plant community?
2. What were the drainage patterns of the San Luis Rey River and Pilgrim Creek? Was Whelan Lake a natural pond prior to the construction of the earthen dam? If not, was the area a major drainage?
3. Was the site ever completely inundated by flood waters? If so, how many times, and in what time periods?

Data Requirements and Methodology

As discussed above, lagoon sedimentation and its effects on local populations has been a topic of concern since the 1960s (Warren et al. 1961; Warren 1964; Quillen et al. 1984; Altschul et al. 1991). Changes in the relative amounts of different shellfish species could reflect the availability of the resource due to increased sedimentation stemming from climatic change (Gallegos 1985, 1987). This hypothesis has, however, been challenged by Hector (1985), who claims that the change in shellfish quantities reflects predation patterns. Faunal analysis will give some idea of the locally available, economically important species. Both on- and off-site pollen samples, microfossil data, and flotation samples will help reconstruct the local floral environment and address questions concerning the lagoon. Geomorphological and hydrological studies will help reconstruct the landscape and address questions concerning flood waters, drainage patterns, and other geological issues.

Chronology

Chronology is perhaps the most fundamental research domain in archaeology. In the San Luis Rey region, as in many locations along the California coast, chronological issues have stymied archaeologists for decades. As discussed in Chapter 3, reconstructing the regional chronology has been difficult for a number of reasons. The paucity of investigated and well-dated sites in the region, archaeologists' attempts to systematize and order a continuous, slow-moving cultural process, a poor understanding of the foraging ability of the early California inhabitants, and confusing terminology have all contributed to the problem of reconstructing the cultural sequence. In addition, sites in the San Luis Rey area tend to be either small specialized loci that do not contain chronometric data or undifferentiated midden sites that were repeatedly used for centuries, if not millennia, and lack stratified deposits demarcating one occupational episode from another.

The culture history of the San Luis Rey area is better known than most coastal areas in southern California due to the relatively large amount of work done in the region. Because no San Dieguito period sites have been discovered in the lower San Luis Rey River basin, the local prehistoric sequence is generally thought to have begun around 5,500 B.C. with the La Jollan Culture. There is a curious gap in the archaeological record between 1,000 B.C. and A.D. 1400. Whether a hiatus in occupation really occurred or the current record is simply a reflection of the paucity of archaeological research cannot be determined at present. Around A.D. 1400 we begin to see moves toward the San Luis Rey Culture which concludes sometime after contact with the Spanish missionaries (circa 1800). Even with this sequence established, major gaps in the framework still exist. The main problems center on the transition periods and cultural differentiation. Determining the age of occupation will help place the site in its regional context by comparing it to contemporaneous sites in the area and will help establish the chronology of the site itself. Relevant research questions are outlined below.

Research Questions

1. When does occupation begin in the San Luis Rey region? Is there any evidence of the San Dieguito Culture?

2. Is the hiatus in occupation between 1000 B.C. and A.D. 1400, noted at SDI-5130 and other major sites in the San Diego region, apparent at SDI-6010, or does this site span the transition?

3. Is a Late Prehistoric period occupation present, and if so, can it be distinguished from a protohistoric occupation?

4. Is there a protohistoric occupation? Was the site occupied contemporaneously with Mission San Luis Rey? When was the final site abandonment?

Data Requirements and Methodology

SDI-6010 represents a large, sparse midden site. It is likely to contain diagnostic artifacts that can be used for relative dating, as well as materials that can be used to provide absolute dates (radiocarbon, obsidian hydration). One objective of the testing phase is to demonstrate that such data exist. We also want to determine whether stratified deposits can be defined or if post-depositional disturbance processes have destroyed the stratigraphy. Stratified deposits would be extremely valuable in addressing questions of culture change or continuity.

Due to the site's favorable location (on a knoll above a good water source, with ample vegetation) and the length of time that humans have been in the area, it is likely that the site has been inhabited a number of times in the past. Stratigraphic analysis will be important in identifying the different occupation episodes. While post-depositional transforms may have mixed the stratigraphy in the upper levels of the deposit, the lower levels may provide the information needed to identify independent occupations. Based on finds in the area (Quillen et al. 1984; Kyle et al. 1988; Gibson et al. 1990) it is anticipated that occupations may date as early as 5,000 - 6,000 B.P. and as late as the historic period.

The reason for adopting an extensive approach is that all dating techniques have inherent problems that render interpreting the results exceedingly difficult. For example, even within one technique, radiocarbon dating, different problems are associated with different types of samples. Dates on shell must consider the reservoir effect of upwelling processes (Taylor et al. 1986; Taylor and Berger 1967; Taylor 1987; Stuiver et al. 1986); those on humates must consider the mean residence time; those on charcoal must consider the problem of "old wood." These problems are often exacerbated at coastal sites where many times one does not find complementary types of samples. We are possibly left, therefore, relying heavily on one organic material, shell, that unfortunately often produces the most problematic dates.

To compensate for the inherent biases associated with different absolute dating techniques, we intend to analyze numerous samples from throughout the site. Ideally, paired samples from the top and bottom of the midden will be analyzed from a number of spatially dispersed test units. There are two assumptions associated with this approach. First, the occupational span represented by the deposit is relatively short. The paired dates, then, should bracket a relatively short period. The samples will be mutually reinforcing, increasing our confidence both in the interpretation and the technique. Disparate dates indicate either contamination or that the assumption is faulty. The second assumption is that the occupational pattern is robust and widespread. Given the number of absolute dates to be analyzed, we reason that, even if local conditions lead to problems with interpreting pairs of dates, overall occupational trends will still emerge.

Temporally sensitive artifacts are to be used as a complimentary line of evidence. We assume that the site represents a short-lived, single-component occupation. We expect, therefore, to find diagnostic artifacts of predominantly one time period. Further, we should find no difference in the age of artifacts between the top and bottom of the midden. Even if the deposit dates to one time period, it is possible that the site represents a seasonal settlement that was occupied repeatedly. In this case, stratified deposits may be found, with cultural strata separated by lenses of sterile soil.

Subsistence and Settlement

Subsistence and settlement have traditionally been two of the favored research domains of archaeologists interested in the California coast. The dynamic nature of coastal environments, along with traceable changes in technology through the prehistoric sequence, has long intrigued archaeologists. In the San Luis Rey area, the contrast between terrestrially adapted (Pauma) sites in the upper drainage and sites focused on lagoonal and coastal resources (La Jolla) near the river's mouth is clear. Whether these sites were part of one large settlement system, or represent distinct systems, and how these trends may have changed over time are major research questions to be addressed. Since the site is one of many recorded in the San Luis Rey River basin, how the site was utilized and how it fit into the regional settlement system are additional subjects of this research domain. Specifically, we want to examine the questions outlined below.

Research Questions

1. What were the components of early Holocene adaptation? How important were aquatic resources? How mobile was the culture? How large was the use territory of individual social units? Was the site occupied permanently or seasonally? If seasonally, what seasons?

2. When did milling stone equipment become important? How did the incorporation of hard seed processing affect the settlement pattern?

3. Were there two settlement systems in the San Luis Rey River Valley, one focused on the upper drainage and another on the lower? Or, was the drainage the locus of one system with two adaptive foci? When did the settlement system(s) crystallize?

4. How did changes in the lagoon affect settlement near the mouth of the river? Can changes in procurement of shellfish, fish, terrestrial animals, and birds be documented, and can these changes be related to shifts in the ecosystem? How was the culture buffered from the vagaries of short-term environmental fluctuations, such as flood events?

Data Requirements and Methodology

To address questions of subsistence and settlement, we need data on diet and the environmental context of the site. The site context can be determined on the basis of geomorphological and microfossil data. The prehistoric diet can be reconstructed from the remains of shellfish and animals, burned plant seeds, and plant microfossils.

Since it is safe to assume that early inhabitants of the project area were hunter-gatherers, it is a reasonable hypothesis that the site was inhabited on a seasonal basis. Determining seasonality will help to confirm or reject this hypothesis. The determination of whether the site was used on a year round or seasonal basis is important in the interpretation of site use patterns. Seasonality will be investigated through vertebrate and invertebrate faunal data, floral resources, and shell analysis. Specifically, representative samples of animal bones, shell remains, and plant microfossils and macrofossils need to be targeted for recovery and analysis.

Lagoon sedimentation and its effects on local populations has been a topic of concern since the 1960s (Warren et al. 1961; Warren 1964; Quillen et al. 1984; Altschul et al. 1991). Changes in the relative amounts of different shellfish species could reflect the availability of the resource due to increased sedimentation stemming from climatic change (Gallegos 1985, 1987). This hypothesis has, however, been challenged by Hector (1985), who claims that the change in shellfish quantities reflects predation patterns. The ability of the site to address subsistence changes depends on the yield of different shell species and the amount of identifiable shells recovered (Botkin 1980).

To link our research questions with the pertinent data requires a design that not only ensures adequate spatial coverage, but also controls for biases inherent in the recovery techniques. Excavation units were placed judgmentally, focusing on areas of suspected features as well as gaps in our overall coverage. The placement of units meets the goal of adequate spatial coverage only if it is coupled with the comprehensive analysis of the various data classes. For certain classes, such as animal bones, this condition is easily met, for the entire collection is to be analyzed. For invertebrate remains, pollen samples, and flotation samples, however, it may not be feasible or necessary to analyze the entire collection. If we find undifferentiated midden, then there may be no reason to suspect a large degree of variability in these data classes. A subsampling design (e.g., selecting shell from a sample of test pits which are themselves a sample of all possible units at the site) can then be devised. As with the case of test pit placement, the selection of test pits from which shell remains were analyzed followed the basic criterion of ensuring adequate spatial coverage from the entire area investigated. Although this design does not guarantee a representative collection, a robust pattern certainly increases our confidence in the results. In addition, if the assemblages are characterized by a high degree of variability either within or between excavation units, we can always increase the size of the sample to be analyzed.

A second methodological issue concerns our recovery techniques. We must ensure that negative results can be interpreted to mean that certain resources were either not utilized or that their residues did not survive in the archaeological record. This goal is extremely hard to meet. In order to achieve it, we must demonstrate that our recovery techniques did not systematically fail to retrieve certain classes of data. This problem is particularly acute in relation to the issues of screen mesh size and the type of screening.

Many coastal animals, particularly fish and waterfowl, are composed of very small bones. If the screen mesh is larger than these bones, then the faunal collection will be biased in the direction of larger animals. Moreover, wet screening techniques generally produce a greater recovery rate than dry screening. Again, this is particularly the case for small bones that may be encased in the soil matrix and are not visible during the screening process.

Our use of 1/8 inch mesh dry screening technique allows direct comparison of the results with most excavations in the region. We recognized, however, that the mesh size and the screening technique potentially biased the results, particularly underrepresenting many lagoonal fish and aquatic bird species. To compensate for this bias, we left a 25 cm x 25 cm soil column in each test pit. Samples corresponding to each level were collected. The samples were floated to recover macrofossils and then screened through 1/32 inch mesh to obtain a sample of remains between this size and 1/8 inch. Although we were not able to analyze the results of these columns as part of this project, the remains have been curated and are available to future reseachers.

Interpretation of subsistence remains from archaeological sites must be made in relation to a regional context. Usually, the frame of reference is the modern environment. However, in highly disturbed areas, such as the San Luis Rey River basin, such an analog is of limited value. An alternative approach is to reconstruct the environment at the time of occupation. Paleoenvironmental data, however, are extremely limited for the San Luis Rey region. For this site, we sought to determine the prevailing environment through the use of off-site pollen samples. As a complementary study we reconstructed the hydrological and physiographic history of the region.

The study of the settlement system will borrow concepts and approaches from site catchment analysis and settlement pattern studies. Environmental data will first be used to reconstruct the prevailing local environment at the time SDI-6010 was occupied. This model will then be coupled with subsistence and technological data to investigate the types of resources actually exploited. The issue of seasonality will be addressed by reference to the seasonal availability of exploited resources. The finding of a wide spectrum of resources available throughout the year will be taken as evidence for permanent occupation. In contrast, if the resources reflected in the various classes of subsistence remains are available during only one part of the year, then a strong case can be made for seasonal occupation.

Chronometric data will then be used to place the site within a temporal framework. Subsistence and environmental data from other contemporaneous sites in the basin will be evaluated to resolve issues of subsistence focus and seasonality. A synchronic settlement model for the time period during which the site was occupied will be created. Models for other time periods will be similarly devised. Diachronic trends in settlement will also be explored for the entire sequence.

Site Structure and Social Organization

Another important research domain concerns the nature of internal site structure and the organizational inferences that can be generated from distributional patterns in features, artifact classes, and other data. Previous studies have demonstrated the usefulness of studying internal site structure. Altschul and Shelley (1987) examined patterns in artifact composition across CA-RIV-269, a late prehistoric-protohistoric period site along Mission Creek in Riverside County. They found that several distinct intra-site areas could be distinguished and that these probably represented activity areas. Specialized work areas associated with production of flaked stone tools, probably by males, and other areas with a generalized suite of male- and female-related activities were identified. Similarly, Basgall and True (1985:5.65) found that thermal features were spatially distinct among Crowder Canyon Sayles Complex sites, also suggesting discrete activity areas. Attempts to identify the size of the group using activity areas and the duration of use have also been made (see Basgall and True 1985:10.18). Freeman and Van Horn (1990:42) suggest that discrete activity areas at the Walker Ranch site represent camp sites used by social segments within Luiseno villages, possibly clans or families. Gibson et al. (1990) found a pattern in the distribution of fire-altered rock at SDI-5130. The units containing large quantities of these rocks were all in the same area that contained "thick deposits of midden, charcoal, debitage, and faunal material" (Gibson et al. 1990:6.23). Specific questions related to the deposit at SDI-6010 are outlined below.

Research Questions

Questions related to site structure and social organization, for SDI-6010 include the following:

1. What are the site's horizontal and vertical spatial boundaries? Are multiple occupations evident?

2. Is artifact distribution across the site uniform, or are there differences in artifact densities that might reflect discrete functional areas within the site?

3. Are features present, and if so, what are their functions? Are features uniformly distributed through space, or are discrete clusters evident? Are discrete midden areas evident?

4. What inferences can be drawn concerning the function of internal site activity areas? Is it possible to determine the size and characteristics of the group using specific activity areas?

5. Are temporal differences evident in site activity areas?

Data Requirements and Methodology

Data requirements for studying site structure include the spatial distribution of artifacts, with respect to their technological, material, morphological, and functional characteristics; the spatial distribution of features and cultural deposits as well as their inferred functions, and temporal data that would determine the periods during which discrete activity areas were used. It is important to determine the degree to which post-depositional disturbance processes and prehistoric behaviors have altered the spatial distribution of artifacts and features. Evaluation of SDI-6010 must consider the potential effect of such processes. Data to address these questions typically derive from controlled surface collections and subsurface recovery. Two aspects of SDI-6010 appear at the outset to influence its ability to answer questions regarding site structure. These are low surface visibility due to current vegetation and historic farming practices that have altered the surface and subsurface distribution of artifacts to an unknown degree. One of the paramount tasks that the testing plan at SDI-6010 must confront is to determine if such practices have affected the integrity of materials to such a degree that further study is unwarranted.

An accurate assessment of the spatial distribution of cultural material is critical to interpreting the archaeological record. An accurate site map showing site boundaries, artifact loci, and historic and prehistoric features will be created through an interpretation of the results from backhoe trenches, test pits, and other relevant information. Soil profiles provide knowledge concerning the vertical array of both cultural and natural material. The maps and profiles aid in the planning of future mitigative work on the site.

Provenience of artifactual material is critical, since the artifacts and their spatial relationships to other materials compose the bulk of the archaeological record (Binford 1962, 1968; Watson et al. 1971, 1984). Formation processes, however, (alluvial action, erosion, road construction, plowing, trampling, etc.) are always acting on the archaeological record and must be taken into account when interpreting the recovered material (Schiffer 1976, 1983, 1987). Since the site is located in an active plow zone, it is anticipated that severe disturbance will be encountered in the upper levels of the deposit. The extent of disturbance will be considered in light of the depth, structure, area, and contents of the site so that the data can be properly evaluated.

Related to the vertical and horizontal extent of the site, is the intrasite relations of materials. The context of cultural remains is important in the analysis of features and activity areas within the site boundaries. While the limits of the testing program prevent a complete mapping of the structure of the site, major activity areas and features will be documented in order to work toward an understanding of cultural patterning.

Exchange and Intergroup Relations

The position of SDI-6010, along the river between the coast and the mountains, would have placed the inhabitants in the middle of any exchange between the major inland and coastal settlements. The potential for studying exchange and intergroup relationships with data from this site is accordingly high. Such studies can potentially inform on aspects of socioeconomic organization in addition to determining resource exploitation territories, intergroup relationships, and exchange processes. Because of its proximity to the Mission San Luis Rey, SDI-6010 also has the potential to address questions concerning the social and economic relationships between indigenous settlements and the early Spanish mission system. Although no data have been recovered from the site that documents an occupation dating to this period, no systematic attempt has yet been made to look for them.

Research Questions

Research questions relevant to this domain are:

1. Are nonlocal items present at SDI-6010? If so, what kinds of items are represented -- utilitarian, ornamental, or ritual-related items; finished goods or raw materials?

2. How do local and nonlocal items differ in inferred function?

3. Is there evidence for on-site production of any goods?

4. What procurement range is indicated by the sources of material goods? What social relationships with groups outside the region are indicated, and with whom are these ties? What is the general direction of presumed exchange (for example, with coastal or mountain populations)?

5. Are temporal components present that would permit the study of change through time in exchange or procurement practices?

6. Is there any evidence of trade or social interaction with the Mission San Luis Rey or local ranchos? What is the nature of the historic artifacts present at the site?

Data Requirements and Methodology

Data requirements for answering these questions include the presence of nonlocal materials and artifacts, such as obsidian, marine shell, various types of stone and shell beads, and steatite, and the inferred functions of artifacts made from such materials. Sourcing of nonlocal materials, most notably obsidian, is also important. Cultural interaction will be measured by the proportional representation of locally available materials as opposed to those that are exotic to the area. Lithic material will comprise the basic elements of analysis, but other categories, such as shell beads, will also be considered. Projectile point styles will be used as another indicator of cultural interaction.

Based on the location of the site and its proximity to the mission, it is a reasonable hypothesis that mission-related artifacts will be found. While this objective is related to other previously stated objectives, specific associations with the mission will help in the interpretation of the site and assist in establishing its associations with other sites in the area.

CONCLUDING REMARKS

Conducting an archaeological excavation is by definition a destructive act. Thus, it is imperative that a well-constructed research design guide the work. One must keep in mind, however, that an archaeological theory simply provides one way of looking at the data and that there are no final truths. In light of this point, in addition to providing possible explanations about the relationship between material culture and past human behavior, archaeology can be seen as providing a view of the society that interprets the data. In other words, similar to a museum reconstruction, the archaeologist provides a window to the past through interpretations from the present (Shanks and Tilley 1987).

Producing a modern text that describes the past is not a neutral act. The evidence recovered depends on the theoretical model from which the archaeologist works, and all interpretation of the data is therefore relative (Fowler 1977; Rowlands 1984). When one examines the reconstruction, certain biases are bound to be evident. It is here that the scientific method in archaeology becomes important. Since the archaeologist maintains hegemony over the past, it is the job of the scientific community to identify and remove biases from the interpretation.

Taking this point of view and applying it to our own research design, illuminates certain aspects of the questions we are asking of the data. Many of the questions concerning chronology center on the controversy of transition periods between certain archaeologically defined cultures. Archaeologists working in southern California tend either to lump "cultures" into broad categories or to split them into vaguely defined "distinct" cultural units. It is important to remember that all archaeological cultures are modern creations. The use of the term culture in large part depends on one's goals. In our case, we are particularly interested in examining human adaptation to the dramatic changes in environmental conditions that characterized the San Luis Rey River basin during the Holocene. Our research, then, focuses on the economic aspects of cultural life. Ideological elements have not been emphasized. Researchers with different orientations to prehistory may view the data from Whelan Lake within a very different cultural paradigm.

Because of our interest in adaptation, we have taken as our subject of inquiry the entire river basin. In so doing, we have encountered a variety of biases in the data that are not easily controlled. At the site level, property ownership precluded an examination of the entire Whelan Lake site. At the regional level, we are constrained by the fact that the database was assembled over time and by different people. The basin has never been systematically surveyed. Thus, we are not in a position to assess the representativeness of the sample of sites that have been recorded. On the other hand, the San Luis Rey River basin is one of the best studied river valleys in southern California, with systematic research having been conducted for nearly 40 years. We are reasonably confident that major trends in settlement can be discerned, with gaps in the data readily defined.

A number of individuals and institutions have been responsible for recording and excavating sites in the San Luis Rey region. A common problem with regional databases is that two investigators will often interpret the same site differently. For example, one archaeologist might describe SDI-6010 as a village site, whereas another might term it a temporary camp. We have dealt with this problem by re-examining the records for all sites in the basin, and deriving an independent classification. We make no claim about the validity of our classification, only that it is internally consistent and replicable.

These critical observations may seem out of place in a research design for test excavations. They are included, however, because they expose possible biases in our interpretations. These biases should not be interpreted as problems specific to this investigation. All scientists must recognize and attempt to factor out bias. To this end, we have tried to present our methodology in a manner such that others can assess the applicability of the data presented herein to future research questions.

CHAPTER 5

FIELD RESULTS

This chapter provides essential background information for later analytical chapters on various classes of artifacts and ecofacts obtained from the Whelan Lake site. Most of the account is taken up by results from field activities conducted at the site. Field work focused on collecting data from the cultural deposits at SDI-6010, in order to address research questions outlined in Chapter 4. To answer the various research questions, four data sets were obtained from the site: artifacts and ecofacts for subsequent analysis, shell and obsidian samples for radiocarbon and hydration dating, pollen and flotation samples for micro- and macrofossil analysis, and stratigraphic profiles and plan maps to provide a control and context for interpretation.

The chapter is divided into two parts. First, a brief description of field proceedings and methods will set the stage for the project's various tasks. The second part of this chapter contains specific information on the findings of trench and test pit excavations, plus a preliminary review of what was observed and/or collected. The soil stratigraphy of the project area is interpreted in the light of geomorphological assessment of the depositional history and geological context of the site. The cultural features identified at the site are then described. Interpretation of these features is partially guided by results of analysis of collected materials and will be continued in later chapters.

PROJECT AREA AND TEST LOCI

The study area included a series of three low knolls flanking the west side of Whelan Lake (Figures 3, 9, 10). Varying densities of prehistoric cultural materials had previously been observed on the surface of the southern two of these elevated areas (Franklin 1988; Franklin and Carrico 1978,1979; Kyle et al. 1988). Only the southernmost knoll was affiliated with a significant artifact density. The surface of the northern knoll had been inspected by SRI personnel prior to the current project. A few ground stone artifacts were observed, so it was decided to include this knoll in the project area. The middle knoll had a low density artifact scatter. After testing by WESTEC in 1988, integrity of this cultural deposit was determined suspect (Kyle et al. 1988; see also Chapter 3). Rather than confining our field work to the main deposit on the southern knoll, we pursued a more comprehensive strategy involving the: (1) assessment of the extent and density of the various artifact scatters, (2) judgemental placement of excavation units within the boundaries of SDI-6010, and (3) placement of additional units in outlying areas containing artifacts.

A rapid start-up of the testing program was hindered by three constraints: (1) partitioned land ownership, (2) dense vegetation, and (3) a subsurface irrigation system. The land ownership problem became apparent when it was found that the project area was divided into three different properties, with parcels owned by the Whelan Lake Bird Sanctuary, the Estate of Ellen Whelan, and the City of Oceanside (see Figure 3). The Whelan Lake Bird Sanctuary owns a large portion of the lake periphery, including the east slopes of the project area's southern and middle knoll and all of the north knoll. Of these areas, the southernmost is the most crucial to the project, as it contains the east portion of SDI-6010. A northwest-southeast oriented dirt road, cutting through the site, serves as a rough boundary, with land between road and lake owned by the bird sanctuary. After initially being granted access to this parcel, it later became clear that this agreement entailed no more than permission to examine the surface. Subsequent requests to conduct subsurface exploration were denied.

Figure 9. Photograph of project area looking south.

Figure 10. Photograph of project area looking northwest from Locus A.

The property owned by the Estate of Ellen Whelan includes the remainder of the south knoll. This lot contains most of SDI-6010. The site location is occasionally used as pasture for the Whelan Lake Dairy and subsurface pipes for a sprinkler system abound. The west half of this knoll was still under cultivation at the time of field work and had been plowed in recent years. Only sporadic shellfish remains were visible in and between the furrows of the field. The City of Oceanside was in the process of buying this property. Permission to cut through the irrigation pipes and crops, and to proceed with the archaeological investigations, which was not immediately granted, was eventually secured.

The City of Oceanside owns a wedge-shaped parcel including most of the project area's middle knoll. The lot contains the various loci of SDI-6010H, located on the knoll's gentle west slope and swale below it (see Schmidt et al. 1992). Surface examination yielded a few prehistoric artifacts. The hill top includes the sparse artifact scatter tested by WESTEC in 1988, interpreted as representing midden fill removed from SDI-6010. Immediate access was available to only this parcel.

The initial plan to define site boundaries through surface inspection had to be abandoned due to the dense vegetation covering the study area. For the same reason, systematic mapping and collection of surface artifacts proved impossible. A cursory survey of the area was carried out, but little information was added to what we knew from previous reports. For easy reference, the remaining area available for subsurface testing was divided into three loci. The east half of the south knoll, containing the main part of SDI-6010, was labeled Locus A. Only the property owned by the Estate of Ellen Whelan was included in the locus. The west half of this knoll, containing the vegetable field, was named Locus B. Proximity to the SDI-6010 surface scatter was decisive in selection of this area for testing. The location of the sparse prehistoric material on the middle knoll was named Locus C. Rather than connoting spatially distinct locales identified at one archaeological site, the above locus designation serves primarily as a convenient tool to group the various proposed trenches.

Three datum points were established in the project area. Datum A (N 994.80/E1020.60) is the site datum, guiding grid layout and the mapping of Loci A and B. Datum B (N1000/E1000) is a control datum, corresponding to a known point, a quarter section corner. Grid coordinates assigned from this point are arbitrary. Datum C (N1050.00/E889.20), located on the middle knoll, was used for mapping the Locus C trenches. Horizontal and vertical proveniences of excavation units, selected artifacts, disturbances, and landmarks were mapped using a transit, stadia rod, and tape.

EXCAVATIONS

In addition to providing controlled samples, subsurface scrutiny provided a more precise verification of both the horizontal and vertical parameters of the Whelan Lake site. Backhoe trenches and/or test pits were excavated in three separate areas, designated Loci A, B, and C (see Figure 3). Of these, only Locus A proved to contain subsurface cultural deposits. Most manually excavated test pits were placed in this locus, with one additional unit placed in Locus B.

Backhoe Trenches

In all, 22 backhoe trenches were excavated. One additional trench was prepared, but then considered superfluous and not excavated. To avoid confusion with the numeric system of trench labeling used for the project's historic component, trenches were given alphabetical designations, ranging from A to W. Trenches were oriented either north-south or east-west and ranged between 15 m and 48 m in length. Width was approximately 70 cm, while depth generally fluctuated between 0.40 m and 1.20 m. Data for all 22 trenches are summarized in Table 7.

Most trenches were concentrated at Locus A, the apparent focus of SDI-6010. Locus B, in the field area just west of Locus A, had practically no cultural material. This locus was tested with a small number of trenches. If prehistoric deposits had been present, additional units would have been placed here. Finally, in Locus C the surface scatter was so sparse that the chance of finding buried deposits was low. WESTEC personnel had previously excavated a number of trenches on this knoll without finding evidence of prehistoric occupation (Kyle et al. 1988). Furthermore, a number of trenches had just been excavated adjacent to Locus C as part of the testing program for SDI-6010H, also without encountering prehistoric deposits. Thus, our trenching activity at Locus C was to be kept to a minimum.

After excavation, trench walls were faced and appropriate profiles drawn. Of the trenches in the midden (Locus A), profiles were recorded for large sections of one of the walls. At the other loci only short, representative sections were drawn. Pollen samples were collected from selected profiles. Diagnostic lithic artifacts, predominantly ground stone tools, were collected from trench walls and back dirt.

Instead of describing each trench separately, the following discussion will consider site (or non-site) aspects as revealed in the trench clusters of the three loci. Of course, trenches at SDI-6010 will receive most of our attention. Locus A comprised 13 trenches (Trenches A-J, L, M, and S), located in or adjacent to what turned out to be the confines of the shell midden. These are the only backhoe trenches that exposed buried cultural deposits. At Locus B five trenches were excavated, namely Trenches K and O-R, located just west of the midden and exposing a rather disturbed area. No midden material was found in these units. At Locus C four units, Trenches T-V, were dug. Also here, no cultural materials were uncovered.

Locus A

Locus A, including most of SDI-6010, consists of a wedge-shaped area, surrounded by three converging dirt roads (Figure 11). The site is situated on a low knoll just west of Whelan Lake and overlooks the San Luis Rey River about 200 m to the southeast. Trenches were placed on the almost flat east portion of this knoll, with several units extending down its southeast slope. Fourteen backhoe excavation units were planned, Trenches A-J, and L-N. Of these, all but one (Trench N) were eventually excavated.

Setting up trench locations was hampered by the dense vegetation, predominantly consisting of tall, non-native grasses. To organize the trenches along a grid system and to allow transit mapping of the units, a swath for a base line had to be cut through the thick shrubbery.

A second problem was formed by the presence of a sprinkler system in the locus. Much of the area just north and west of Whelan Lake is now used for cattle grazing, and a network of irrigation pipes and sprinklers has been put in to maintain a sufficient grass cover. In Locus A, 18 sprinkler heads occur, encased in concrete footers spaced 15 m to 20 m apart and interconnected by subsurface PVC pipes. A main line valve is located in the northern portion of the locus. No clear idea of exactly how the sprinklers were connected could be obtained, and thus no way of laying out trenches that would avoid irrigation pipes could be devised. It was decided not to let the location of trenches be influenced by the lay-out of the visible portion of this system. As this locus was of crucial importance to the project, a number of the pipes had to be cut. Care was taken, however, to insure that the main line would be avoided.

Table 7. Summary of Data for All Backhoe Trenches in the Project Area.

Trench Number	Locus	Length (m)	Orientation	Wall Profiled	Midden depth (m)		Shell Density			Associated Test Pits	Feature	Pollen Samples Collected	Disturbance	Comments
					max	min.	west	center	east					
A	A	38	E-W	N,S	0.70	0.30	L	H	M	6				
B	A	25	E-W	N	0.90	0.50	L	H	H	7			Trash pit in west end	MA**
C	A	25	E-W	N,S	0.60	0.40	M	H	H	10	4			MA**
D	A	37	E-W	N	1.00	0.50	L	H	H	9			Trash pit in west end	
E	A	35	E-W	N,S	0.50	0.30	H	H	L	11	3		Irrigation pipes	
F	A	37	E-W	N	0.55	0.35	-	M	H				Trash pit in west end, irrigation pipes	MA**
G	A	30	E-W	S	0.90	0.30	H	H	H	2, 12	1, 2	From Feature 1	Irrigation pipes	MA**
H	A	48	E-W	N	0.80	0.00		L	H				Irrigation pipes (one along N wall), Trash pit in west end	
I	A	20	E-W	N	0.60	0.00		L	L				Irrigation pipe	
J	A	18	E-W	N	-	-	-	-	-					
K	B	35	N-S	W	-	-	-	-	-				Plow zone	
L	A	30	N-S	E	-	-	-	-	-					
M	A	30	N-S	E	-	-	-	-	-					
N*	A	-	-	-	-	-	-	-	-					
O	B	18	N-S	E	-	-	-	-	-				Plow zone	
P	B	18	N-S	W	-	-	-	-	-				Plow zone	
Q	B	18	N-S	W	-	-	-	-	-				Plow zone	
R	B	18	N-S	W	-	-	-	-	-				Plow zone	
S	A	18	N-S	W	-	-	-	-	-			Control sample		
T	C	35	E-W	N	-	-	-	-	-					
U	C	35	E-W	N	-	-	-	-	-					
V	C	35	E-W	S	-	-	-	-	-			Control sample		
W	C	25	E-W	S	-	-	-	-	-					

*This trench was proposed but not excavated

**Moist area with high clay concentration

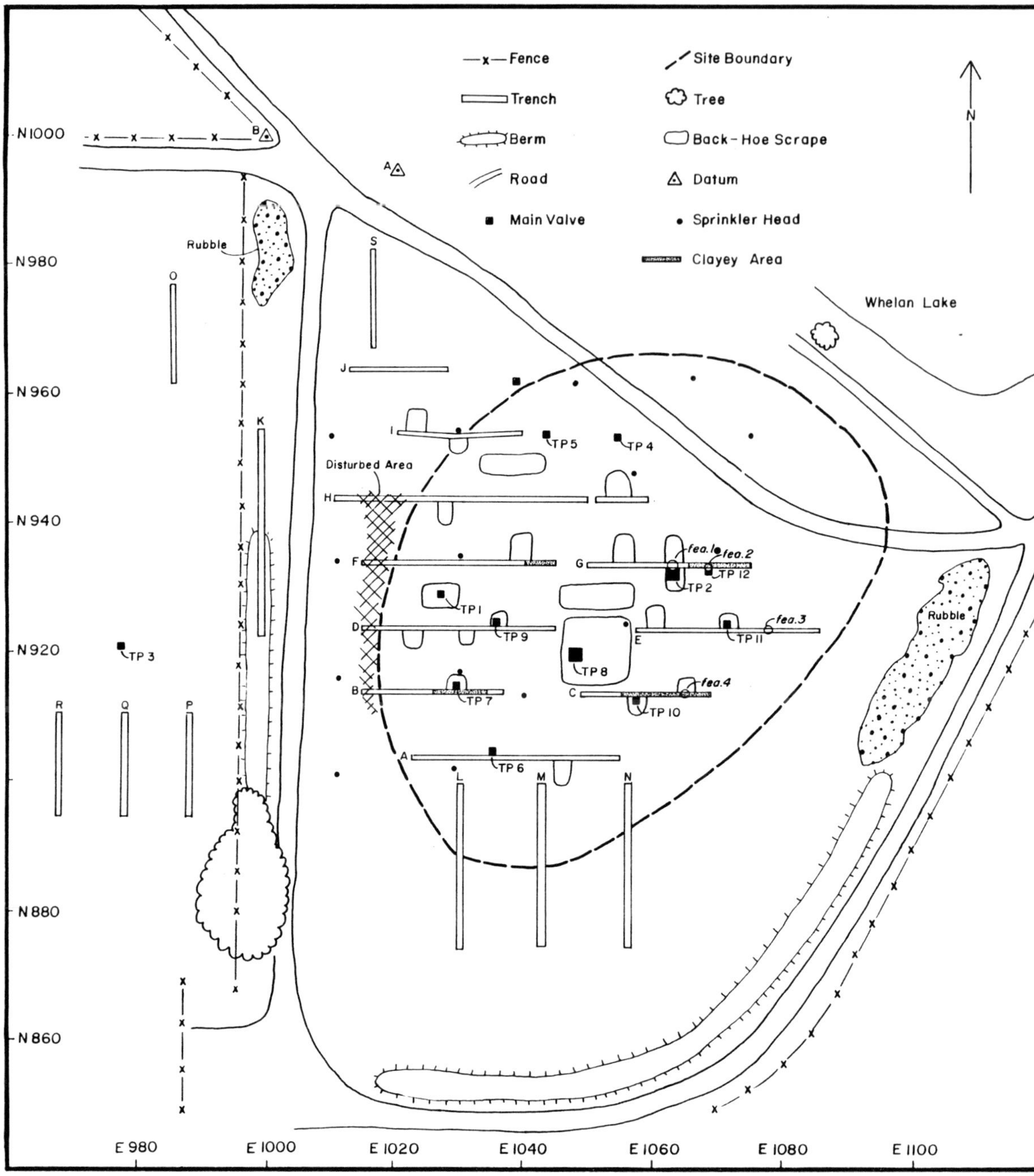

Figure 11. Map of CA-SDI-6010, the Whelan Lake site, showing Loci A and B.

A third problem arose when the scheduled time for trenching arrived, but permission to cut through the irrigation pipes was not yet obtained. Trench excavation in the field area of Locus B was also on hold, forcing the initiation of trenches in Locus C, the area of least interest. Fortunately, backhoe access to the more relevant area of the southern knoll was obtained shortly thereafter.

First, 10 east-west oriented backhoe trenches, labeled A-J, were excavated. All but one (Trench J) exposed shell midden deposits. Then four more trenches, all oriented north-south, were laid out. Trenches L-N of this latter group were placed so as to define the south boundary of the midden. Of these southern three trenches, only Trenches L and M were excavated. In these it was noted that the midden ended about half-way through each unit. The fourth unit, Trench S, was placed in the northernmost part of the locus. In this trench, only a few shellfish remains were encountered, documenting that the midden does not extend beyond Trench J.

Trenches at Locus A were generally spaced at 10 m intervals, with lengths ranging between 15 m and 48 m. Through relatively tight spacing and an emphasis on long units it was hoped to obtain an optimal understanding of midden stratigraphy and generate sufficient potential to expose cultural features.

With the group of 13 trenches at Locus A, most of the shell midden constituting the Whelan Lake site was securely defined. Only the northeast boundary remained imprecise as no excavation was possible in the lot owned by the bird sanctuary. An approximate location of the site's northeast edge was later established by careful surface inspection of the slope towards the lake.

The trench profiles of Locus A display a noticeable variation in site-specific attributes. These attributes include: (a) soil horizons, (b) archaeological features, (c) artifact density and diversification, (d) modification of midden soil, and (e) disturbances. In the following, variability within the midden as seen in the various trench profiles will be discussed. Most of the account will be incorporated within a summary of the stratigraphic context of the midden. A digest of disturbances noted, critical for determination of site integrity, follows the archaeological assessment of the shell midden.

Site Stratigraphy

Observations of profiles in the trenches excavated at SDI-6010 indicate the presence of a cultural layer, characterized by organically enriched soils, whole shells and shell debris, bone fragments, and lithic artifacts. The shell midden is approximately oval in form, its long axis extending 90 m from northeast to southwest, its short axis about 60 m wide. The midden fill is up to 1 m thick at its center but thins rapidly towards the margins of the site.

Table 7 demonstrates that, although variations exist between individual trenches, the composition of the midden is fairly homogeneous across the site, and the stratigraphy is uniform throughout the locus. Three soil horizons (A, B, and C), dividing into five strata, were identified in the trench profiles.

Stratum I. A 15 cm to 20 cm thick mantle, consisting of colluvial and possibly aeolian deposits mixed with midden fill, covers the entire site. The matrix of this A horizon is a dark brown (Munsell 10YR 3/3, dry) to very dark grayish brown (Munsell 10YR 3/2, dry), soft, sandy loam. Shell and lithic artifacts occur, but are not as plentiful as in Stratum II below. The stratum is pervaded by an abundance of rootlets and other organic matter; rodent burrows are also common. Bioturbation, irrigation, and possibly plowing, have all contributed to a degree of mixing between Stratum I and the upper portion of Stratum II. Although soil color is similar for the two strata, texture and consistency are different, and the bottom of Stratum I is easy to identify in spite of the mixing.

Stratum II. Stratum II is the 20 cm to 100 cm thick anthropic fill of the shell midden. Like Stratum I, the fill is classified as an A horizon. The matrix is primarily a dark brown (Munsell 10YR 3/3, dry) to dark yellowish brown (Munsell 10YR 3/4, dry), hard, coarse sandy loam. Varying densities of cultural materials occur throughout this stratum. Most of it is shell, with lesser amounts of lithics and animal bone. Occasional small charcoal fragments occur. Bioturbation and human agencies have impacted portions of this stratum, but overall, archaeological integrity is fair. No plow zone was identified in the midden fill.

Table 7 and Figure 12 compare midden depth across the site. Recorded depths include the disturbed upper A horizon, Stratum I. Stratum II reaches a maximum depth in the east end of Trench D, close to the site center. Rather than a gradual and even thinning of the midden towards its horizontal confines, a high degree of fluctuation in depositional depth is apparent. This is demonstrated in the west end of Trench E, also near the site center. Here, 12 m east from the reading in Trench D, midden depth has decreased to 0.30 m. Overall, the thickest cultural deposits are in the west part of the site, as indicated by maxima of 0.70 m in Trench A, 0.90 m in Trench B, 1.00 m in Trench D, 0.80 m in Trench H, and 0.60 m in Trench I. In the east half of the site, the midden is no thicker than 0.50 m. Figure 12, a diagram of approximate midden depth across the site, further exemplifies how deposition mass is centered in the west half of SDI-6010.

Trenching defined most of the site's south and northwest boundaries (see Figure 11). On the southeast slope of the knoll, profiles of Trenches L and M reflect how the midden becomes gradually thinner, ceasing south of N 888.50 (Figure 13). Similarly, to the west, midden fill vanishes at E 1030.50 in Trench I, at E 1024.60 in Trench H, and at E 1020.60 in Trench F. Precise north and east boundaries were not obtained from trench profiles, and a gap occurs in the southwest site perimeter. The northeast portion of the site is owned by the bird sanctuary and was not available for trenching. A segment of the west boundary has been obliterated by a deep north-south oriented borrow pit, refilled with asphalt and rubble. In spite of these lacunae, sufficient information was available to reconstruct the circumference of the shell midden.

Four cultural features were identified in the trenches (see Table 7). All originate in Stratum II and protrude into the underlying B horizon. In profile, only the fill of Feature 1 is distinct, largely composed of fire-cracked rocks (FCR) and ash. Conversely, the fill of Features 2, 3, and 4 appears practically identical to that of the surrounding midden, only distinguished by an occasional FCR and some small charcoal flecks.

Table 7 highlights shell density as observed in the trench profiles. Highest shell density was recorded in the center of the midden. The site core is roughly oval in shape, measuring 56 m x 38 m and oriented along the same northeast-southwest axis as the midden itself (see Figure 11). Within this area, shell density is high to moderate. Outside the core, density drops off to low and very low. In the heart of the midden, six discrete areas of greater density can be singled out.

It is intriguing to note that four of the areas of greater shell density coincide with four enigmatic moist, clayey pockets in the midden exposed in 8-m- to 12-m-long sections of Trenches B, C, F, and G (see Figure 11). All four of these moisture-retaining areas displayed an intensified shell density. Sporadic charcoal flecking (not observed in the general midden fill) was noted in three of the areas. The moist soil is a very dark brown (Munsell 10YR 2/2, moist), organically enriched, loose, silty clay loam. No doubt due to easy digging in the soft matrix, rodent burrowing is abundant. Features 2 and 4 are associated with two of the moist areas.

<u>Stratum III</u>. Underneath most of the midden (but not outside it), the top 10 cm to 30 cm of the B horizon has been enriched with calcareous material. Classified as a Bk horizon, the stratum is mostly a white (Munsell 5YR 8/1, dry) and extremely hard calcium carbonate. Descending into the B horizon, the caliche gradually disappears, leaving no abrupt boundary between Strata III and IV. The Bk horizon is most conspicuous in the heart of the midden. Rather than representing a natural stratum, the caliche is interpreted as shell residue, leached down into the B horizon. Strangely, it is absent under the moist clay area found in Trench B. Otherwise, it is present in all of the midden core and in much of the surrounding area of lower shell density.

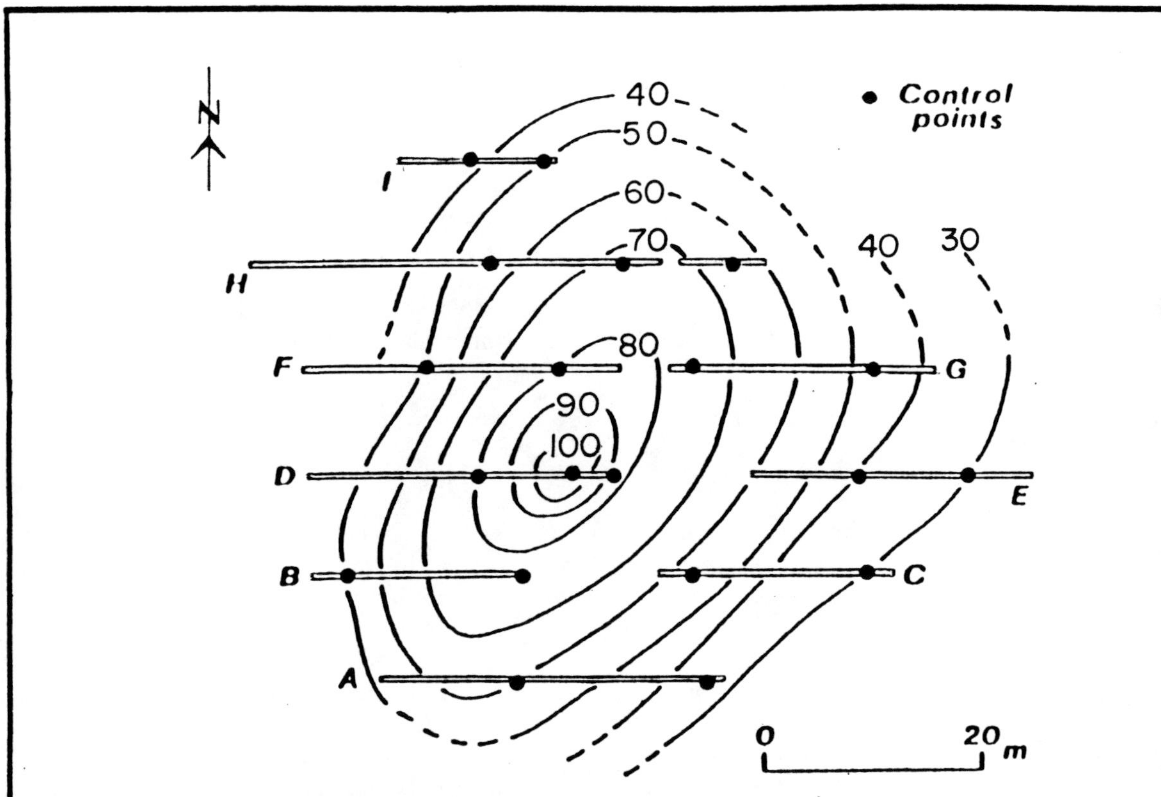

Letters refer to trenches
Contours indicate midden thickness in centimeters, measured at control points

Figure 12. Diagram showing depth of shell midden across SDI-6010 (by Antony R. Orme).

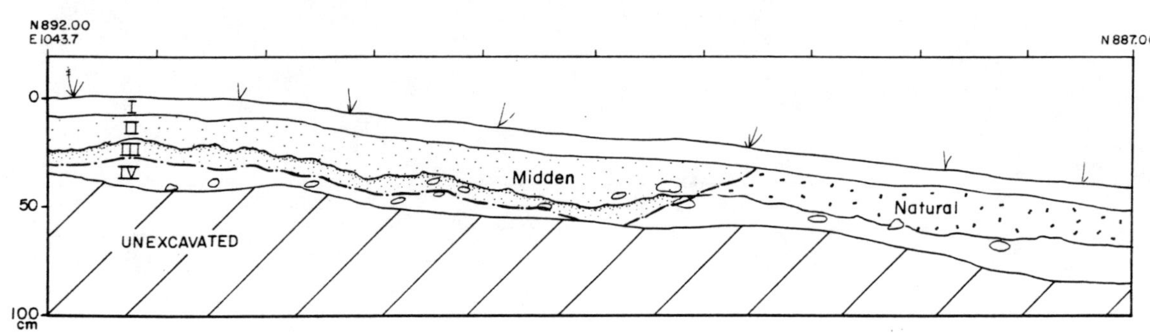

Figure 13. East profile of Trench M, showing south boundary of the shell midden (SDI-6010).

Stratum IV. This is the lower, non-calcic portion of the B horizon, occurring throughout the site. Stratum IV ranges between 20 cm and 50 cm in thickness and is a brown (Munsell 10YR 5/3, dry) loamy to clayey, coarse sand with no artifacts.

Stratum V. A basal stratum composed of poorly sorted sands, gravels, and pebbles of a Late Pleistocene river terrace deposit underlies the site. On the east slope of the knoll, the stratum was found at 0.40 m depth in the east end of Trench A. Here, the stones are of pebble and cobble size and include granite rocks in varying stages of decay. To reach this stratum near the site center, a pit was dug in Trench G to a depth of 1.25 m. The stratum is classified as a C horizon and described as a dark yellowish brown (Munsell 10YR 4/4, dry), very coarse sand, intermixed with pebbles and cobbles.

Disturbances

Three types of disturbances were noted in the Locus A trenches: (1) irrigation activities throughout the midden, (2) borrow/dumping pit excavation at the west edge of the site, and (3) bioturbation throughout the midden. A fourth disturbance to the site, outside of the locus, is formed by a dirt road cutting through the site's east portion.

As expected, given the presence of sprinklers on the surface, an underground system of irrigation pipes was exposed. PVC pipes, placed horizontally at a depth ranging between 60 cm and 70 cm, were found to connect with short vertical pipes at the various sprinkler locations. A number of trenches cut through one or more pipes. Much of the north wall of Trench H has a pipe running along its profile (Figure 14). Trench I had to be curved northward to avoid a concrete-cased sprinkler head. Disturbance associated with the sprinkler system is twofold. First is the impact to the site caused by excavation for pipe burial. The impact from the slit trenches for these pipes is surprisingly low. Apparently, the character of the midden fill is of sufficient homogeneity to leave no visible traces of soil displacement after rapid backfilling of trenches that, in most cases, were probably not much wider than the 1 1/4 inch diameter of the pipes themselves. In contrast, pipe trenches dug into the B horizon left highly visible traces. For example, in Trench E, two 2-inch PVC pipes intersect the trench at a depth between 70 cm and 90 cm. In both cases, Stratum III was penetrated and material from Stratum II was used to fill the disturbance. Similarly, the (4 inch) pipe running along the north wall of Trench H has a visible disturbance above it. Above this pipe, the profile shows only a low density of shell, whereas the opposing trench has abundant shell in the corresponding location.

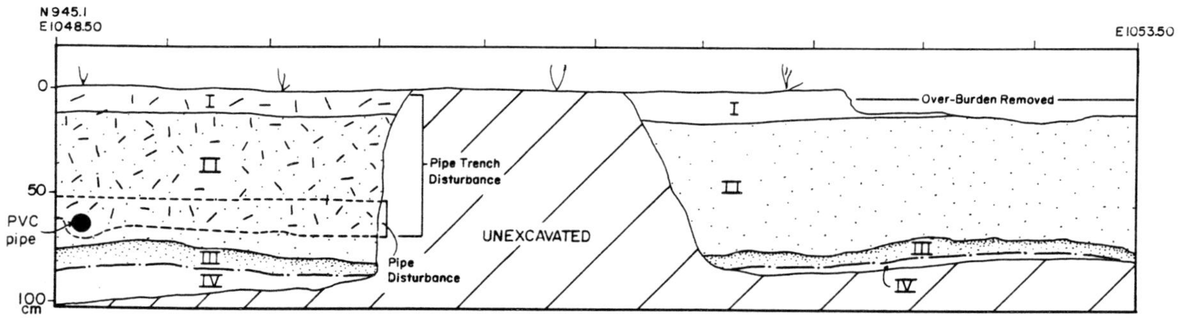

Figure 14. North profile of Trench H, showing irrigation pipe disturbance.

A second impact to the site, also associated with the sprinkler system, consists of a downwards leaching of cultural materials by irrigation water. Even though not directly visible, this process has undoubtedly obliterated some of the "finer" stratigraphy, washing out evidence of prehistoric activities and more recent cultivation.

At the west edge of the midden, a deep, disturbed area was noted in the west portions of Trenches B, D, F, and H. Here, asphalt and other debris have been dumped in what appears to be a pit over 2 m deep in some areas and encompassing an area about 30 m long (N-S) and 10 m wide. The pit appears to stop at 3 m to 5 m from the dirt road along the west edge of the site. The disturbance has clipped off the west edge of the site. Shell density in the trench portions bordering on the disturbed area is fairly low and the impacted site area is probably small. This may be the location from which midden fill was "exported" to Locus C in modern times (see Kyle et al. 1988).

As has been noted before, bioturbation, especially rodent burrowing, is fairly extensive in some areas. Most of it occurred in the soft soil displaying moisture retention.

Locus B

Located on the same knoll as Locus A, this locus is just west of SDI-6010. The two loci are divided by a north-south oriented dirt road. Except for a narrow strip of land between this dirt road and a barbed wire fence, all of Locus B was in use as an agricultural field (see Figure 10). Until recently, this field had been well-maintained, as demonstrated by orderly furrows and a still-functioning system of irrigation pipes and hoses located on the field surface. During field work, vegetable crops were still present, although in various stages of decay.

Five north-south oriented backhoe trenches were excavated in Locus B. Trench K was placed outside the field in the narrow band to the east of the barbed wire fence. Trenches O-R were located in the vegetable patch. Stratigraphy of all trenches was markedly dissimilar to that of the Locus A trenches. Not more than slight variations were noted between the stratigraphy of Trench K and that of the other four trenches placed in the field area (Figures 15 and 16). Because of the lack of cultural material found in these trenches, no further trenching was done in Locus B.

The profile of Trench K displays a 15 cm thick top stratum, composed of a dark grayish brown (Munsell 10YR 4/2, dry), sandy loam. Below this stratum, to a depth of about 35 cm, a layer of a grayish brown (Munsell 10YR 5/2, dry) coarse sandy loam was exposed. This stratum, an apparent mixture of B horizon and a darker soil, has evidence of heavy churning and is interpreted as a plow zone. Without a clear break, the plow zone fades into the underlying B horizon, a brown (Munsell 10YR 5/3, dry) loamy, coarse sand. This third stratum extends to an undetermined depth. No shell or other cultural material was observed in this trench.

The O-R group of trenches uniformly has a top stratum, 3 cm to 8 cm thick, of a brown to dark brown (Munsell 10YR 4/3, dry) sandy loam. Underneath this, to a depth of 25 cm to 45 cm, a plow zone occurs of a very dark grayish brown (Munsell 10YR 3/2, dry), sandy loam. Within this plow zone, two to three superimposed substrata with minor color and texture differences can be identified. Most likely, these substrata represent different plowing episodes. The plow zone fades into the unmixed B horizon, a brown (Munsell 10YR 5/3, dry) loamy, coarse sand. In Trenches O and Q an underlying Bt horizon is visible in the profile. This basal stratum is a dark yellowish brown (Munsell 10YR 4/4,dry) sandy clay loam.

The absence of prehistoric deposits in these trenches indicates that the Whelan Lake site does not extend into Locus B. Plowing activities have occurred in Locus B, and top soil appears imported for agricultural purposes. While of no direct importance to the archaeology of SDI-6010, the Locus B profiles help in assessing whether that site was ever plowed. Located on the same knoll as SDI-6010, all five trenches of Locus B show a vaguely stratified plow zone, probably reflecting several churning episodes. No such layering is evident in the midden fill of Locus A, indicating that a plow zone is not present at the Whelan Lake site.

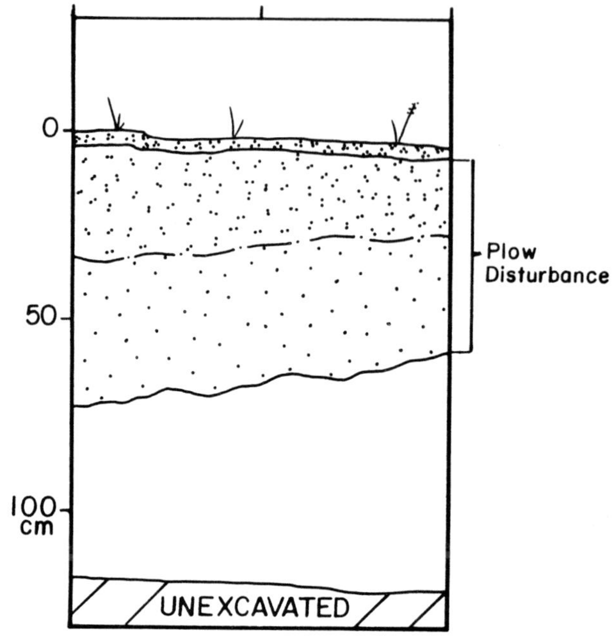

Figure 15. One meter section of west profile of Trench K.

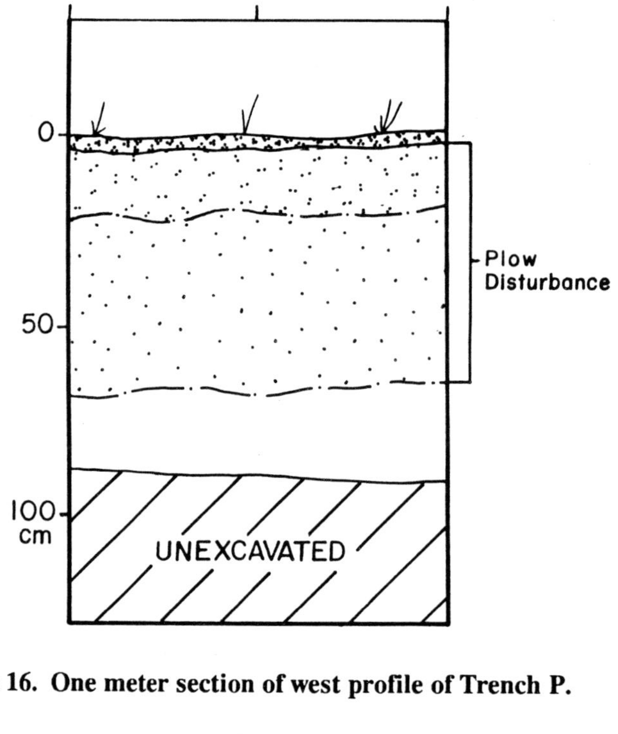

Figure 16. One meter section of west profile of Trench P.

Locus C

Four trenches (Trenches T-W) were excavated in Locus C, all oriented east-west and ranging between 20 m and 35 m in length. Most were located north of the trenches excavated by WESTEC on this knoll. No cultural deposits were encountered, confirming the findings of WESTEC.

Trenches T-V had a similar stratigraphy to Locus A (Figure 17). Stratum I, from 0 cm to 3 cm, is the A horizon, a brown (Munsell 10YR 5/3, dry) crumbly, loose clay loam. Stratum II, from 3 cm to about 35 cm, is the B horizon, a brown (Munsell 10YR 5/3, dry) hard sandy clay with massive structure. Stratum III, from 35 cm to an undetermined depth, is the Bt horizon, a dark brown (Munsell 10YR 3/3, dry), extremely hard silty clay.

Trench W, the northernmost unit of the four and located at a slightly higher elevation, had a somewhat different stratigraphy (Figure 18). Stratum I is the 3 cm to 10 cm thick A horizon, distinguished by a very dark grayish brown (Munsell 10YR 3/2, dry) sandy clay with crumbly structure and a loose consistency. Stratum II, from 10 cm to about 55 cm depth, is the Bt horizon. This stratum consists of a very dark gray (Munsell 10YR 3/1, dry), hard, silty clay with a massive structure. Stratum III, from 55 cm to an unknown depth, is a Btk horizon, a white (Munsell 5YR 8/1, dry) silty clay containing calcareous materials. The composition of this natural Btk horizon is dissimilar from the cultural Bk horizon occurring at SDI-6010. The difference is that the calcic horizon at the shell midden vanished gradually downwards in the B horizon, whereas the Btk horizon observed at Locus C appears to be a solid and thick layer of caliche, not confined to a thin coat on the B horizon.

Test Pits

Twelve test pits were hand excavated and screened during the project (see Figure 11). Ten of the test pits were 1 m x 1 m units and two were 2 m x 2 m units. Almost 8 cubic meters were excavated. Eleven test pits were placed within the confines of SDI-6010. One was excavated outside the site boundary in Locus B, the field area to the west of the shell midden. Test pit stratigraphy mirrored and further defined that described for the trenches. Table 8 summarizes grid coordinates, size, depth, and volume for each test pit.

Goals and Methods

The fill of all test pits was screened through 1/8 inch mesh hardware cloth. Unfortunately, water screening was not possible. All artifacts and ecofacts were collected. In most of the midden, the hard, compact soil formed a considerable constraint on excavation speed.

Because of delays in obtaining permission to conduct mechanical excavations at Locus A, three of the test units, Test Pits 3, 4, and 5, were dug before any backhoe trenches were excavated. Of these, Test Pit 3 was the only unit placed in Locus B. Located outside the site confines in a deep plow zone, this unit was terminated at 50 cm depth. Test Pits 4 and 5 were placed in the northeast portion of SDI-6010. All test pits at SDI-6010 were excavated down to non-cultural soil. In Test Pit 8, a 2 m x 2 m unit, only the northwest quarter was excavated to the B horizon. Test Pits 2 and 12, each containing a feature, were excavated as standard test pits from the surface until the top of the features were encountered. Thereafter excavations were confined to the feature fill.

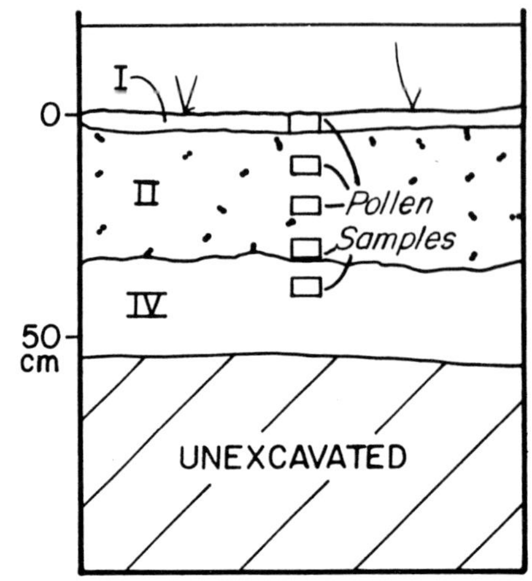

Figure 17. One meter section of south profile of Trench V.

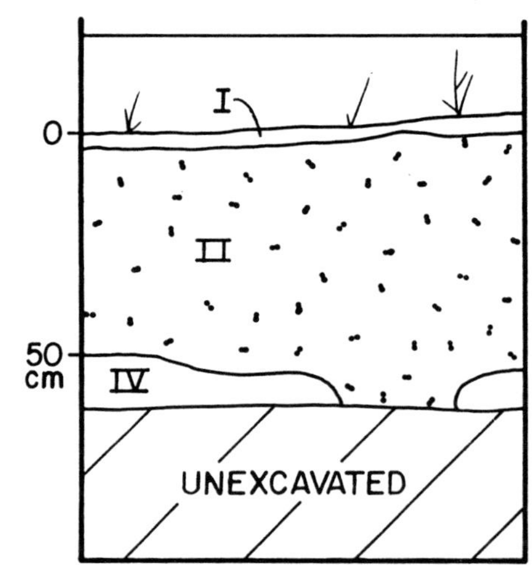

Figure 18. One meter section of south profile of Trench W.

Table 8. Surface Area, Depth, and Volume of Test Pits.

Test Pit Number	Grid Coordinate	Surface Area (sq m)	Depth (m)	Volume (cu m)	Comments
1	N930.80 E1028.80	1	0.45	0.45	
2	N938.50 E1062.35	4	0.55	0.98**	Only upper 0.20 m of total unit was excavated. Below 0.20 m depth, excavation was restricted to feature fill.
3	N919.60 E977.50	1	0.5	0.5	
4	N953.80 E1053.85	1	0.5	0.5	
5	N934.00 E1043.85	1	0.7	0.7	
6	N905.10 E1035.50	1	0.65	0.65	
7	N915.10 E1029.90	1	0.7	0.7	
8	N920.45 E1047.80	4	0.7	1.60**	Only northwest 1x1 m square was excavated to 0.70 m depth Rest of unit was terminated at 0.30 m depth.
9	N925.10 E1036.40	1	0.7	0.7	
10	N913.50 E1056.55	1	0.45	0.45	
11	N925.10 E1073.20	1	0.3	0.3	
12	N932.50 E1062.35	1	0.65	0.4**	Only upper 0.15 m of the total unit was excavated. Below this depth excavation was confined to feature fill
Totals		18		7.93	

*Tests pits 4-6 excavation included the disturbed upper 10-15 cm of the A horizon (Stratum I). For all other tests pits, this stratum was removed by backhoe or hand prior to excavation and is not included in the test pit depth.

**Volume represents an adjusted value.

Obtaining controlled artifact samples of sufficient size for subsequent analysis was a major goal of test pit excavation. Thus, most units were judgementally placed in areas of high artifact density, either as seen in a specific segment of a trench profile (Test Pits 6, 9, and 10) or on the surface (Test Pits 4, 5, and 8).

A second important object of test pit excavation was to define and sample features identified in the trenches. Of the four features detected, Features 1 and 2 appeared to retain the greatest integrity, and therefore were excavated. Some of the four clayey, organically enriched areas exposed in the trenches, also needed testing. Accordingly, Test Pits 7 and 10 were placed in the moist areas uncovered in Trenches B and C. Test Pit 12 sampled the moist area exposed in Trench G, simultaneously excavating Feature 2.

Finally, a third goal of putting in test pits was to identify possible intrasite variability. To provide a workable context for spatial analysis, obvious gaps in test pit distribution across the site were adjusted by being "filled up" with Test pits 1 and 11, even though cultural materials appeared sparse in these locations.

Of the 12 test units, Test Pits 3, 4, 5, and 6 included the relatively more disturbed upper stratum of the A horizon (Stratum I); elsewhere this stratum was stripped prior to test pit excavation (see Figure 11). To facilitate manual excavation, 19 areas of varying sizes were scraped off by the backhoe front loader. Of these scraped areas, seven were eventually selected for test pit placement. No collections were made from these stripped areas.

Test Pit Evaluation

Table 9 summarizes artifact counts and ecofact NISP per level for each test pit. Comparison of these data allows us to fine-tune our initial reconstruction of site composition gleaned from the Locus A trench profiles. An in-depth evaluation of spatial differences in distribution of the various artifact and ecofact classes across the site follows in Chapter 11. At this point, it is useful to offer a brief synopsis of contrasts and similarities between the various test pits. In doing so, we will be able to guide the various analyses of artifacts and ecofacts, at the same time preparing a prelude to an interpretive examination of site composition in Chapter 11.

The three primary classes of material collected from the test pits are shellfish remains, animal bones, and lithic artifacts. One small fragment of human bone was found. Occurring in general midden fill, it was recovered from Test Pit 8. Two ceramics were recovered from the site; one was found on the surface, and the other sherd came from Level 1 of Test Pit 8. Additional samples collected include charcoal, macrofossils, and microfossils.

Of all the units, Test Pits 2, 8, 9, and 10, contained most cultural material. Because Test Pits 2 and 8 were partially excavated as 2 m x 2 m units, artifact counts and NISP calculations need to be adjusted (i.e., counts of cultural material per cubic meter). After adjustment, Test Pits 2, 8, 9, and 10 still rank highest on the scale of general artifact/ecofact density per test pit. Apparently, the four test pits represent locations of intensified use within the midden. This comes as no surprise, as Test Pit 2 was placed over a hearth (Feature 1), whereas Test Pits 6, 8, and 10 were all excavated in areas chosen for their high densities of shell. Furthermore, Test Pit 8 is located in the exact center of the midden.

Test Pits 1, 4, and 5 rank lowest on the scale of artifact/ecofact density per unit. This was to be expected, as all three units are located outside the heart of the midden (see Figure 11).

In Test Pit 2 it was noted that while the upper two levels contained few *Donax* shell, substantial amounts were found in the fill of the underlying feature, a hearth. Thus, an obvious example of intrasite ecofact distribution was identified. One of the goals of the shell analysis will be to see if intrasite variation extended beyond the Test Pit 2/Feature 1 contrast (see Chapter 8). Similarly, both lithic and faunal analyses will consider spatial distribution of material across the site (see Chapters 7 and 9).

Table 9. Artifact Counts and Ecofact NISP per Level for Each Test Pit.

Unit	Level	Area (sq m)	Depth (m)	Volume (cu m)	Chipped Stone	Ground Stone	Bone	Shell
Test Pit 1		1	0.45	0.45				
	1				2		2	7
	2				5		1	
	3				4		5	13
	4				2		14	6
	5						2	
Test Pit 2		4	0.55	0.98				
	1				15		22	60
	2				18	5	37	148
	3				53		76	260
	4						1	69
Test Pit 3		1	0.5	0.5				
	1				9			
	2				2			
Test Pit 4		1	0.5	0.5				
	1				4		1	8
	2				2	1	1	20
	3				4			28
	4				8		2	52
	5				1		1	18
Test Pit 5		1	0.7	0.7				
	1							16
	2				1			
	3				4		1	
	4				2			12
	5				1			8
	6				4		2	30
	7				1			27
	8				2		2	9
Test Pit 6		1	0.65	0.65				
	1				13		1	12
	2				17		6	82
	3				11		4	94
	4				4		7	60
	5				11		18	79
	6				2		2	26
	7						3	10
Test Pit 7		1	0.7	0.7				
	1				3		4	32
	2				2			20
	3				4		15	102
	4				4		9	144
	5				3		2	117
	6				1		1	37
Test Pit 8		4	0.7	1.60				
	1				24		83	138
	2				60		428	204
	3				83	1	276	228
	4				18		68	38
	5				25		61	47
	6				12	1	31	118
	7				9		19	54
Test Pit 9		1	0.7	0.7				
	1				5		2	8
	2				10		10	29
	3				20			
	4				17		17	
	5				28			45
	6				2		4	30
	7				5		4	19
Test Pit 10		1	0.45	0.45				
	1				19		7	43
	2				17		17	104
	3				13		24	127
	4				10		18	57
	5				11		14	28
Test Pit 11		1	0.3	0.3				
	1				17		15	28
	2				20		18	78
	3				6		12	
Test Pit 12		1	0.65	0.4				
	1				25	3	19	237
	2				16		21	90
Totals					691	11	1410	3356

Clayey Areas in the Midden Fill

As discussed above, four areas were found in the midden fill where the soil is clayey, has a high organic content, and retains a great deal of moisture. Features 2 and 4 are each associated with such an area. Test Pits 7, 10, and 12 were excavated in three of the clayey areas. Table 9 shows that the NISP counts for shell from these test pits confirm that the moist, clayey areas are associated with high shell densities. Furthermore, charcoal, uncommon in the general midden area, was recorded for all three test pits.

Moisture retention of these areas can be explained by the high clay content, but where did the clay come from and what was the cultural association? Some degree of loam is present in the general midden fill and the underlying B horizon, but no easy explanation is available for clay concentrations in these areas of apparently intensified cultural activity.

The following optional scenarios for clay importation or formation appear worth considering:

(1) Clay formed in situ:
  a) Organic colloids formed in a natural depression
  b) Mineral clay formed by physical or chemical weathering
(2) Clay was brought in, adhering to imported shell
(3) Clay was brought in, suspended in water used for food processing
(4) Clay was brought in for a specific purpose:
  a) Creation of a floor or other activity surface
  b) Deposited historically, possibly dredged out of Whelan Lake.

As to 1a, no depressions associated with the clayey areas can be discerned in the site's current or paleotopography, canceling this option. Similarly, 1b has to be discarded, as a much longer time span is needed for clay formation than suggested by the A horizon context of our examples.

Options 2 and 3 both have potential to explain the clayey areas. Much of the shell occurring at CA-SDI-6010 originates from mudflats in the mouth or lagoon of the San Luis Rey River. These mudflats are partly composed of clay from river sedimentation. Thus, there is a good chance that together with the shell, adhering clay was deposited in the site's processing areas. Likewise, importation of clay suspended in water brought in for food (shell) processing might be a possibility. The San Luis Rey River, not more than 200 m away, probably represented the primary water source for the site. Varying quantities of clay particles would be present in this water. When used for cooking purposes, the clay may have been removed from the water through decanting, and deposited in the processing areas.

Option 4a may likewise be possible. Not much is known about the types of dwellings or structures associated with shell middens. Structures once present on the midden may have contained clay as a building material, now surviving in the form of large clayey pockets in the midden.

Finally, option 4b, clay deposits resulting from mud dredged out of the lake, seems unlikely. This option does not explain why these areas coincide with greater densities of cultural materials.

In sum, without conclusive evidence, the clay areas can only be linked to a set of optional explanations. Of these, the ones associating the areas with specific prehistoric activities stand the best chance of surviving future scrutiny. Only through additional work at the site, such as soil analysis and clay source determination, can the various hypotheses of clay importation be evaluated.

Feature Descriptions

Four archaeological features were identified at the Whelan Lake site. All were detected in walls of backhoe trenches. Two of the features were partially excavated: Feature 1 with Test Pit 2, and Feature 2 with Test Pit 12. Features 3 and 4 were recorded, but not excavated. A description of all four features follows.

Feature 1

Excavation Strategy

Feature 1 was exposed in both walls of Trench G. Best defined in the south wall, it appeared as a lens of fire-cracked rocks (FCR) overlying a deposit of ashy material (Figure 19). The rocks formed a pocket in the midden fill, but the ashy layer was embedded in the sterile B horizon. In the south wall of the trench, the feature occurred over a 1.40 m length (from E 1062.65 to E 1064.05). In the north profile it was only visible over a 0.90 m exposure (from E 1062.90 to E 1063.80).

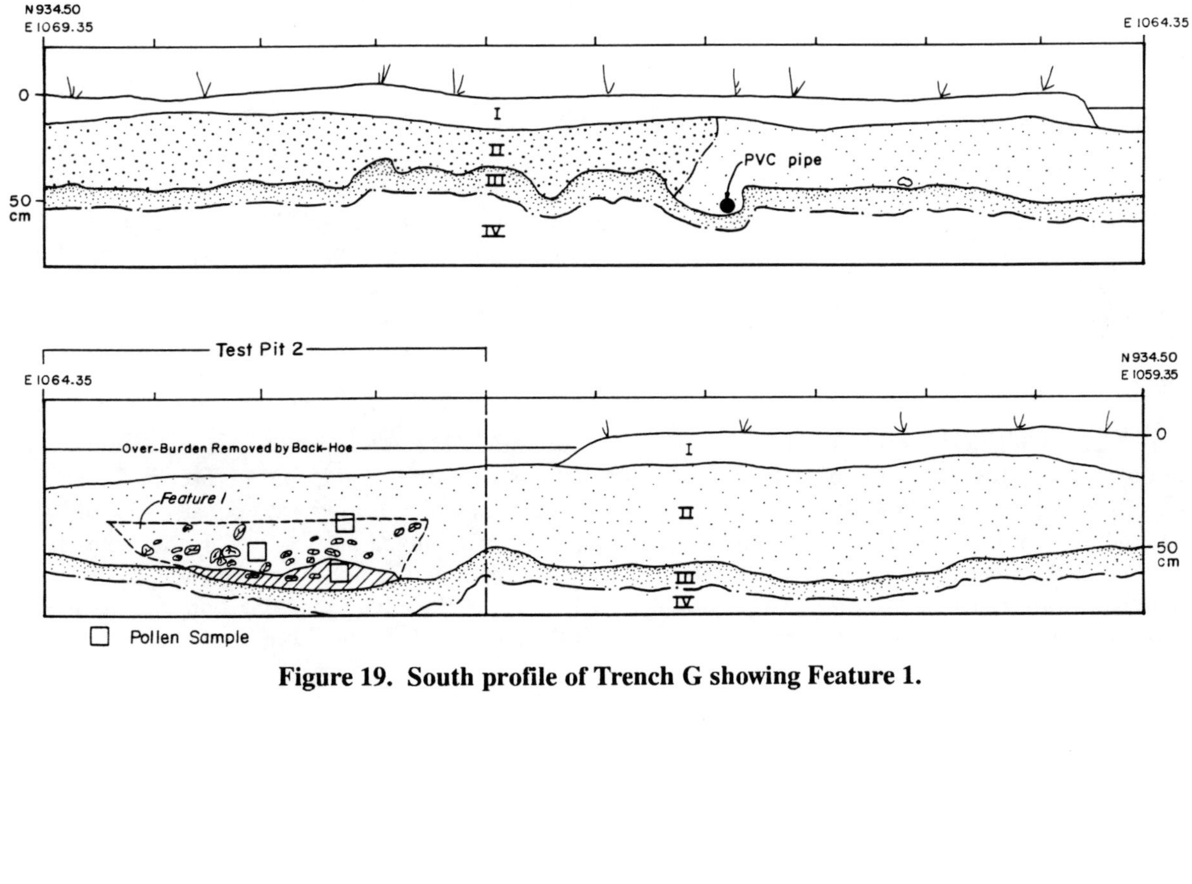

Figure 19. South profile of Trench G showing Feature 1.

Most of Feature 1 was located south of the trench, and this portion was excavated. After a backhoe had removed 18 cm of disturbed A horizon, Test Pit 2 was set up over the feature. The test pit was a 2 m x 2 m unit with a datum at N 932.50/E 1062.35. All elevations were measured from a stake, located just west of the unit and positioned so that 0-elevation corresponded to the local surface before stripping.

Four levels were excavated, of which Levels 1-3 were each 10 cm thick. For practical reasons, Level 4 was allowed to reach a thickness of 18 cm. Levels 1 and 2 consisted of midden fill above the feature. The top of the feature, denoted by a lens of FCR, was encountered at the top of Level 3 at 37 cm depth. Thereafter, excavation was confined to the feature fill defined by a semi-circular lens of FCR. The feature wall above the B horizon was poorly delineated, with definition largely depending on the contrast between FCR fill and rock-free midden. The feature bottom was reached in Level 4, at 65 cm depth.

Description

The feature consists of a circular pit, originating in the lower portion of the midden and protruding into the B horizon (Figures 20 and 21). Diameter at the top of the FCR fill is about 1.40 m. Pit depth (from top to bottom) is about 30 cm. The wall of the depression in the B horizon is fairly irregular, yet sloping down gradually towards the center of the feature. Pit diameter at the boundary between A and B horizons is about 90 cm. The pit has a flared, bowl shape, with a wall that is steepest at its west end.

Stratigraphy

In the south profile of Trench G, Feature 1 stratigraphy is well-illustrated. Two distinct strata are present in the feature fill. The top fill, from 37 cm to 60 cm depth, is a wavy, disc-like layer of FCR encased in midden fill. The matrix is composed of a dark grayish brown (Munsell 10YR 4/2, dry) coarse sandy loam. Underneath this lens, from 52 cm to 65 cm depth, is the ashy bottom fill of the pit. This fill is composed of a dark gray (Munsell 10YR 4/1, dry), fine sandy ash containing rocks, shell fish remains, and numerous small charcoal fragments.

Midden fill overlies and partially surrounds the feature. The feature matrix is not completely uniform across the site as to color, texture, and artifact content. In this segment of Trench G, it is a dark brown (Munsell 10YR 3/3, dry), coarse sandy loam, containing a moderate density of artifacts. The adjacent midden/B horizon boundary undulates, but not as much as observed in the southeast portion of the site. As noted in most trench profiles in the midden, the upper part of the B horizon is enriched with calcium carbonate, presumably leached from the shell contained in the midden. Occurring as a compact coat of calcareous material on the surface of this horizon, it gradually vanishes further downwards.

Figure 20. Photograph of Feature 1 as seen in Test Pit 2, Level 3.

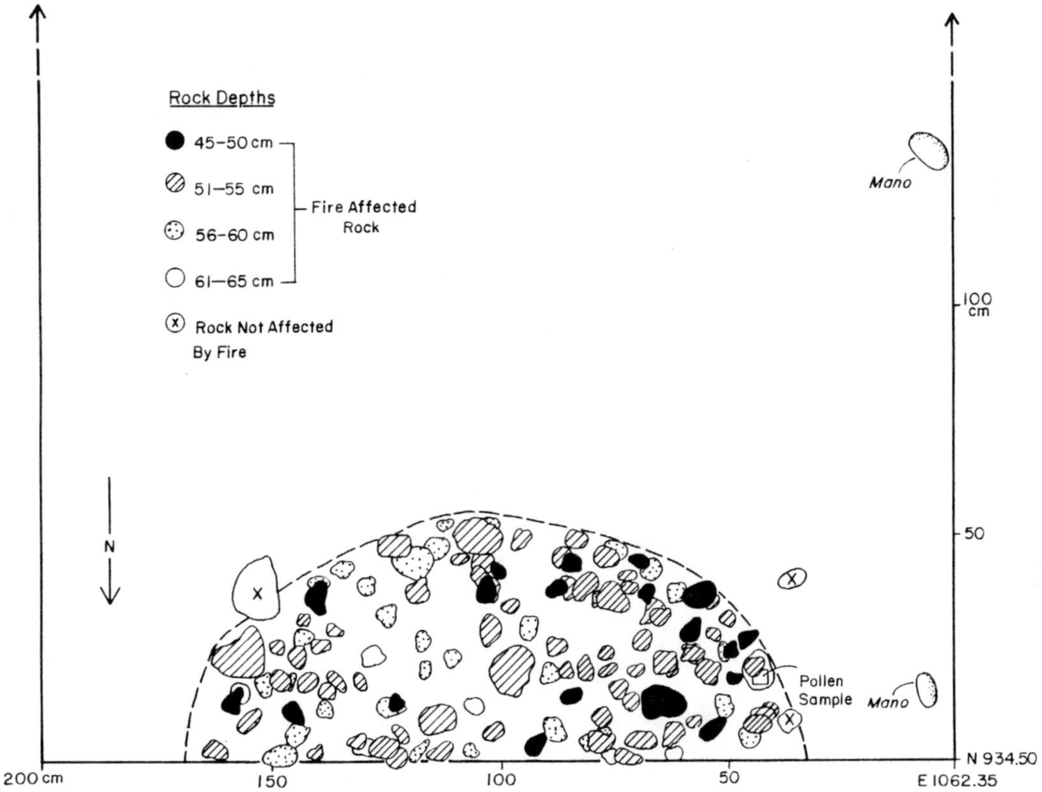

Figure 21. Plan map of Test Pit 2, showing Feature 1.

Feature Fill and Artifacts

The thermally altered rocks in the feature fill were mostly concentrated in the area between 52 m and 60 cm in depth. Only a few stones occurred in the lower 5 cm of the feature. The west half of the pit had the highest density of rocks (see Figure 21).

Shell within the feature differs from that found in the surrounding midden fill. Levels 1 and 2 (above feature fill) hold large quantities of fragmented shell, mostly *Chione* and *Argopecten* and no *Donax*. Levels 3 and 4 (feature fill), on the other hand, contained a high concentration of intact shell, with *Donax* predominating.

No chipped stone or ground stone was encountered in the feature fill. The absence of lithic material in the fill is remarkable, especially when contrasted with the abundance of small flakes recovered from the test pit in the area above and adjacent to the feature.

Two well-shaped manos were found to the west and southwest of the feature. Both were placed upright, one at 45 cm depth and the other at 50 cm depth (see Figure 21). Such manos are uncommon in the site's artifact assemblage. The two manos appear to be associated with the feature and the depth at which they occurred may correspond to the activity surface during the time that the feature was in use. Several ground stone fragments were also collected from the area surrounding the feature.

Interpretation

The feature is interpreted as a hearth, used for food processing. The presence of specific shell species in the feature fill suggests that shellfish preparation may have occurred in the feature. Plant processing may also have occurred in the hearth. Analysis of macro- and microfossil samples collected from the feature may help identify some of these plants (see Chapters 10 and 11).

Several burned fragments of animal bone were collected from Feature 1, indicating that the hearth was used for meat preparation. Dog (*Canis* sp.), hare (*Lepus* sp.), pocket mouse (*Pergonathus* sp.), and rabbit (*Silvilagus* sp.), were identified. In addition, one burned bone fragment of an unidentified bird species was present.

Feature 2

Excavation Strategy

Feature 2 was encountered in Trench G, about 8 m east of Feature 1. Exposed in the south wall only, Feature 2 occurred over 1.60 m length (from E 1072.35 to E 1073.95) (Figure 22). The feature is situated in one of the four areas found in the midden where the soil is clayey and holds moisture. Only after the soil of the trench profile had dried was the feature identified.

Feature 2 was partially excavated in Test Pit 12, a 1 m x 1 m unit placed alongside the south wall of the trench. This was the last test pit dug during the project, and at this point, a time constraint had come into play. Prior to excavation, the disturbed upper 15 cm of the A horizon was stripped from the feature top. The test pit datum was located at N 933.50/E 1072.35. The elevation datum corresponded to the unstripped surface above the test pit. It was decided to excavate the feature in only two levels.

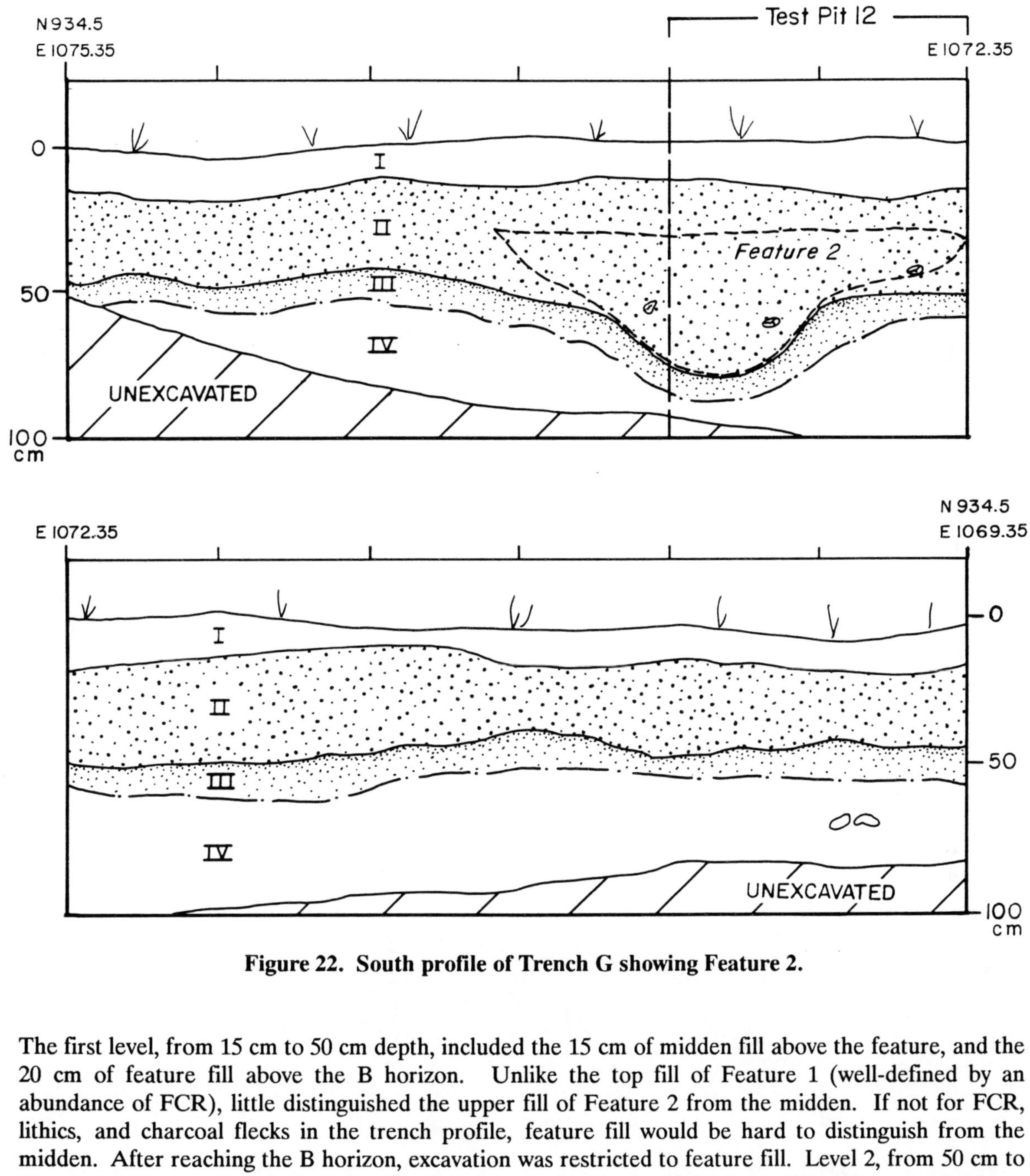

Figure 22. South profile of Trench G showing Feature 2.

The first level, from 15 cm to 50 cm depth, included the 15 cm of midden fill above the feature, and the 20 cm of feature fill above the B horizon. Unlike the top fill of Feature 1 (well-defined by an abundance of FCR), little distinguished the upper fill of Feature 2 from the midden. If not for FCR, lithics, and charcoal flecks in the trench profile, feature fill would be hard to distinguish from the midden. After reaching the B horizon, excavation was restricted to feature fill. Level 2, from 50 cm to 77 cm depth, corresponded to the fill of the depression in the B horizon. Flotation samples were collected from both levels, and pollen samples were taken from the fill remaining in the south profile of the test pit.

Description

The feature was a roughly circular pit. Diameter (reconstructed) at the top of the B horizon is approximately 1.50 m. The center of the feature occurs in the northeast corner of the test pit

(Figure 23). Location of pit bottom is accentuated by an increased downwards slope of the feature wall.

Feature Fill and Artifacts

The fill was still moist when excavated. Throughout the feature, the fill was a homogeneous, very dark grayish brown (Munsell 10YR 3/2, moist), clayey, sandy loam. This matrix is similar to that of the surrounding moisture-retaining portion of the midden. Artifacts include shell, lithics, faunal material, and charcoal. No ashy bottom layer, such as present in Feature 1, was noted.

A low quantity of FCR was encountered. A small, dense pocket of rocks was found about 10 cm above the B horizon, at the west edge of the feature. Most other FCR was evenly distributed through the fill.

Interpretation

Charcoal and FCR in the fill identify the feature as a hearth. Interestingly, much of the FCR was located in a small cluster on the edge of the feature. Thermally altered stones in a hearth context can be interpreted in two ways. First, after heating in a fire, the rocks may have been put in a container to bring water to a boil. Meat, shellfish, or plants could be cooked in this manner (see Kroeber 1925).

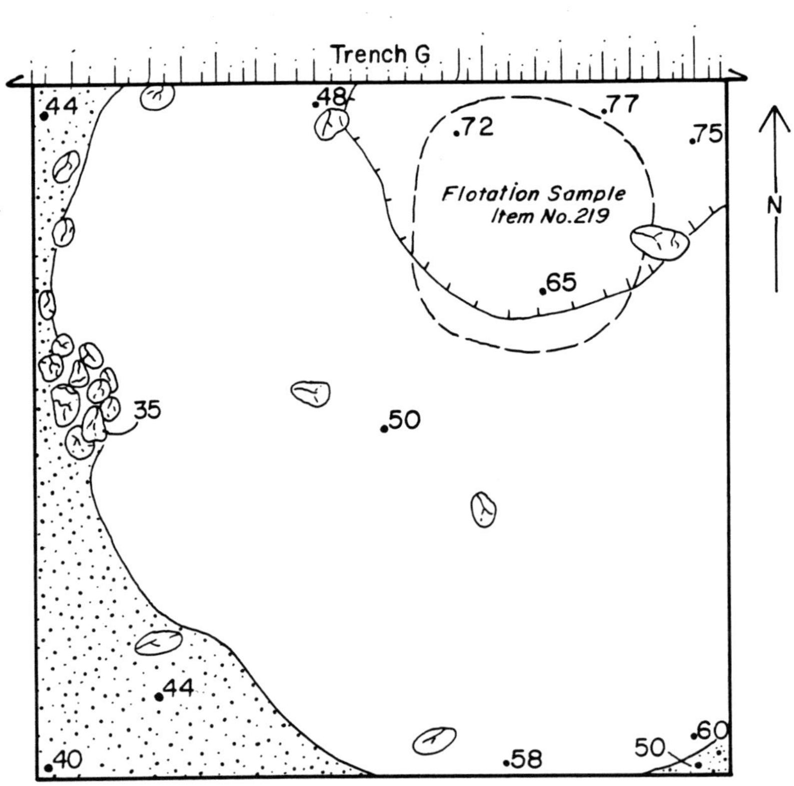

Figure 23. Plan map of Test Pit 12, base of Level 2, Feature 2.

Secondly, stones may have been placed on top of burning coals to form a roasting context for specific types of food. In California, this kind of food preparation was most common in a desert environment, but also occurred on the coast, primarily associated with agave or yucca. In both cases, the hearth warranted periodic cleaning. Whereas Feature 1 still seemed to have a complete fill of rocks, rocks are sparse in Feature 2, possibly the result of cleaning or maybe indicating a functional variance.

Compared to Feature 1, animal bone is slightly less common. Hare and rabbit are the only dietarily important species identified. Similarly to Feature 1, the fill of Feature 2 contains a slightly higher percentage of *Donax* than the general midden fill.

Feature 3

Feature 3 was found in the east end of Trench E, exposed in both walls. In the north profile the feature occurs over a length of 1.10 m (from E 1081.90 to E 1083.00); in the south profile there is a 0.65 m exposure (from E 1081.70 to E 1082.35). The feature was not excavated. It appears as a depression in the B horizon, with a fill that is near-identical to that of the midden itself (Figure 24). If not for some charcoal flecks observed in profile, both in and above the depression, feature designation might even be questionable. A fire-affected pebble is seen in the north profile, resting on the bottom of the feature. The midden is fairly shallow in the proximity of the feature, ranging between 20 cm and 40 cm in depth. Shell density is low compared to the west half of the trench.

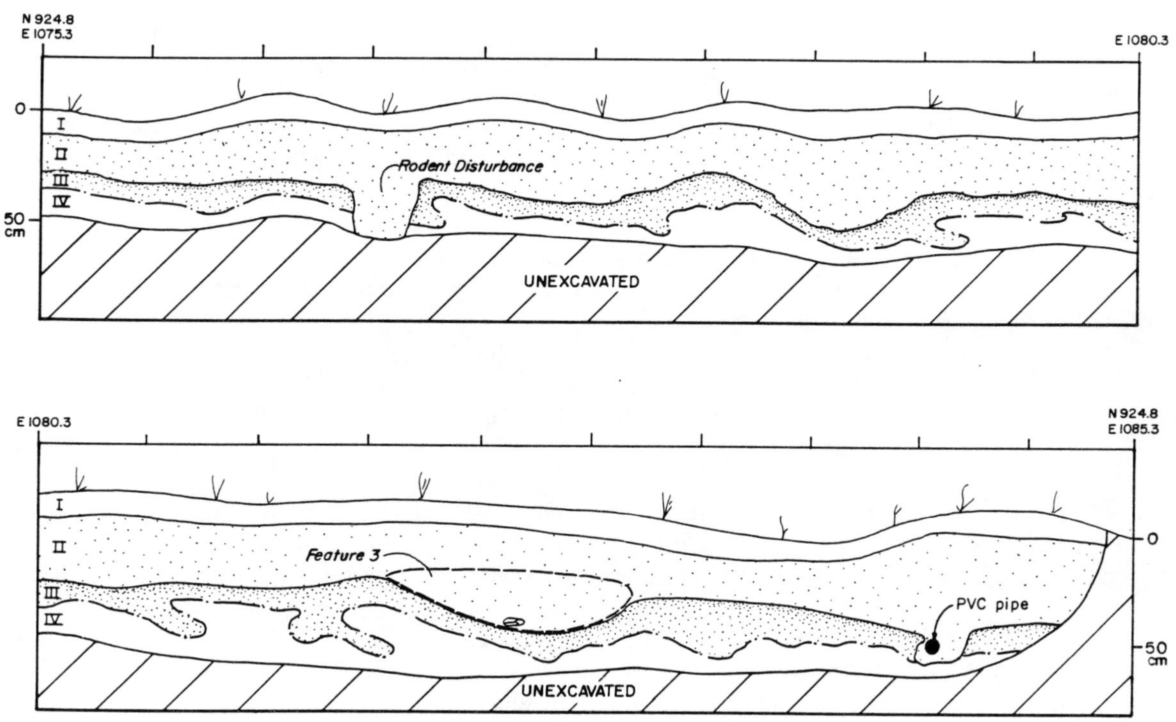

Figure 24. Feature 3, displayed in the north profile of Trench E.

Feature 3 is located on the edge of the midden core where shell density drops off rapidly (see Figure 11). The boundary between the midden fill and the underlying B horizon undulates considerably in this portion of Trench E. Some of the deep and narrow dips in the B horizon are obviously rodent related, others are associated with the PVC pipe irrigation network. Some of the wider and shallower depressions visible in profile may be remnants of prehistoric activities. Of these depressions, only Feature 3 had a fill that was reasonably distinct from the surrounding midden.

All that can be ascertained about the feature is that it consists of an about 20 cm deep pit in the B horizon with a few charcoal fragments and an FCR in the fill. The charcoal occurs to about 10 cm above the pit wall, suggesting that the feature originates somewhere in the midden fill, unless rodents brought the material upwards. The feature is interpreted as a hearth.

Feature 4

Feature 4 was found in the east portion of Trench C (Figure 25). In the north and south wall it was observed over a 0.70 m exposure (from E 1063.70 to E 1064.40). Apparently, the trench cut the feature in half. Feature shape appears to be circular, with a diameter of 0.75 m. Feature 4 was not excavated or sampled and all observations come from the trench profiles.

Analogous to Features 1 to 3, Feature 4 appears as a shallow but pronounced depression in the B horizon. In the south profile of the trench, an FCR is visible in the fill. The north profile shows two more FCRs in the midden fill just east of the feature. Except for several miniscule charcoal fragments, feature fill matrix is like that of the midden.

The feature is situated within one of the four areas identified at SDI-6010 where the soil is clayey and organic, and capable of retaining a high degree of moisture. Midden depth near the feature ranges between 25 cm and 40 cm, thinning towards the east end of the site. Rodent disturbance is extensive, adding to the undulating character of the B horizon. A relatively wide and shallow pit occurs in the north profile, just west of the feature. Nothing in the fill sets this adjacent depression apart from the surrounding midden.

Feature shape, charcoal in the fill, associated FCR, all suggest that the pit functioned as a hearth. The few FCR noted east of the feature most likely originate from it and possibly occur on the ancient surface associated with the hearth.

Summary and Conclusions

The testing program has confirmed that the Whelan Lake site is a shell midden, confined to the eastern portion of a low knoll overlooking the San Luis Rey River to the southeast. Mechanical and manual excavations in two other loci with sparse surface scatters of cultural material yielded no buried deposits. From backhoe trenches and test pit excavations, a satisfactory understanding of the stratigraphy of the site has been obtained. Within the midden, backhoe trenches exposed four cultural features and four areas distinguished by clayey, moisture-retaining soil. All features are hearths, whereas the more enigmatic clayey areas are tentatively interpreted as locations associated with shellfish processing. A sufficient number of test pits was put in to obtain data required to address the project's research questions. Besides further fine-tuning the site's stratigraphy, test pits helped explore and sample two of the features and three of the clayey areas, at the same time allowing generous sampling of the general midden area.

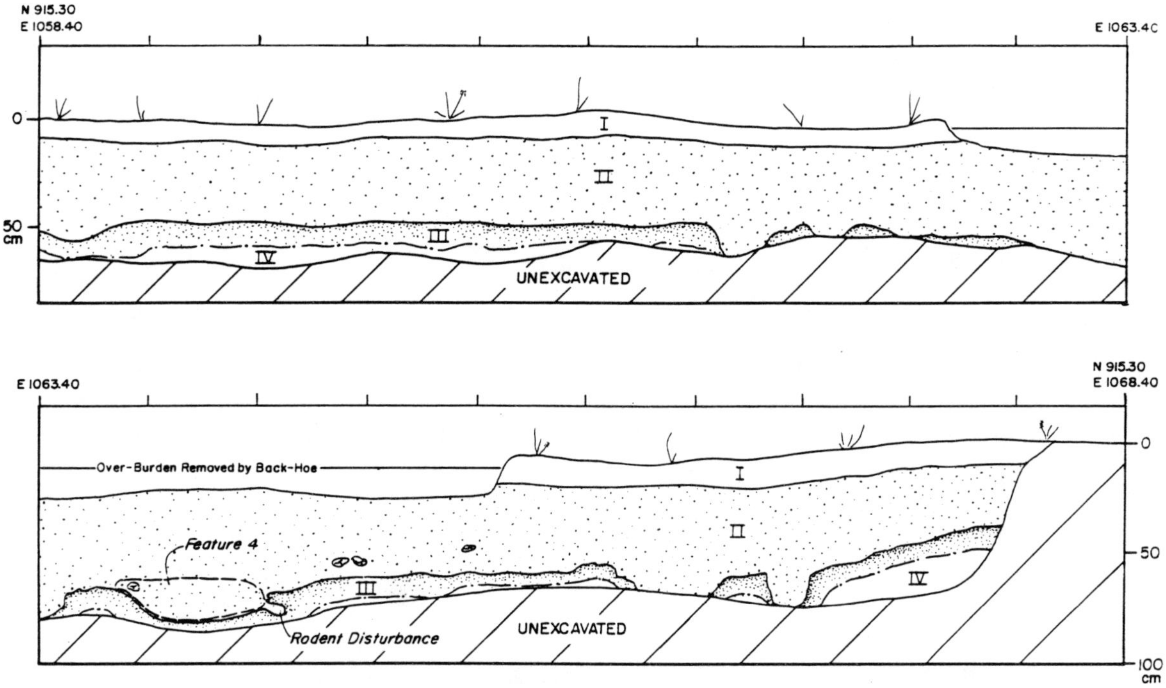

Figure 25. North profile of Trench C, showing Feature 4 and moisture retaining area.

Shell samples collected from the test pits were submitted for radiocarbon dating, and obsidian specimens were subjected to trace element analysis and hydration band measurement. Chronometric analysis is discussed in Chapter 6. The lithic materials, shellfish remains, and animal bone collected from the test pits will be discussed in Chapters 7, 8, and 9, respectively. Flotation and pollen samples were taken from selected locations in test pits and trenches. Results from macro- and microfossil analysis are detailed in Chapter 10.

CHAPTER 6
CHRONOMETRIC ANALYSIS

The Whelan Lake site yielded two types of samples that were subjected to chronometric analysis: shell for radiocarbon dating and obsidian for hydration measurement. Unfortunately, very little charcoal was recovered from the site. None of the charcoal samples exceeded 1 g, and the few samples that came close to this amount were recovered from visibly disturbed areas; therefore, no charcoal samples were submitted for radiocarbon dating.

RADIOCARBON DETERMINATIONS FROM SHELL SAMPLES

Certain biases are associated with the composition and decay of radioactive carbon in shell. Because of the importance of absolute dating for estimating the time the Whelan Lake site was occupied, it is important to review some of the factors that must be accounted for in interpreting the dates. Shell is comprised of an organic matrix (referred to as conchiolin) which incorporates inorganic salts, primarily carbonate and lesser quantities of magnesium carbonate and calcium phosphate (Taylor 1987:49). Because conchiolin comprises less than a few percent of shell, virtually all of the ^{14}C determinations are based on the inorganic fraction. Marine shell can yield reasonably accurate dates if the *reservoir effect* is taken into account to factor out the effect of carbonates "upwelled" from deeper sections of the ocean. This correction factor varies from one segment of coastline to the next, depending on the degree of mixing between near surface and "upwelled" carbonates. As suggested by Robinson and Thompson (1981:48) for southern California, Freeman (1991) used a correction factor of 690 ± 35 years. The 690 ± 35 year correction factor, however, pertains to marine shell, not estuarine shell, such as the *Chione* dated from the Whelan Lake site.

No studies have been conducted to determine what the specific correction factor is for estuarine versus marine shells along the southern California coast. However, in a recent study comparing the radiocarbon age of *Rangia cuneata* (a brackish water species) with nearby marine shell along the Texas coast, *Rangia cuneata* was found to yield a date between 250 and 300 years younger than its marine counterpart (Sam Valastro, Director of University of Texas-Austin Radiocarbon Laboratory, personal communication 1991). Because of the similar habitats of *Rangia cuneata* and *Chione* sp., it is reasonable to subtract 275 years for the reservoir effect.

Another factor that has a bearing on radiocarbon determinations for shell is the ^{13}C adjustment for isotopic fractionation. "Fractionation" refers to the differential absorption rates for certain radioactive isotopes (Freeman 1991:2). Marine shell is thought to be especially prone to modifications in isotopic absorption, with the result being a younger ^{13}C age than the actual age of the shell. In Freeman's (1991:4, Table 1) review of 19 radiocarbon dates of marine shell obtained from five sites along the Ballona Escarpment, the ^{13}C adjustment was 421.6 ± 32.9 years. Freeman (1991:4) reported that the combined isotopic and reservoir effect corrected datings were 268.3 years younger than the corrected ^{14}C dates.

With ^{13}C and reservoir effect for estuarine shell adjustments applied in tandem, a near cancellation of the adjustments occurs. Because the reservoir effect correction factor for estuarine shell is estimated at approximately 275 years, the expected difference between the uncorrected and the combined isotopic and reservoir dates appears to be negligible, only about a 7-year difference. Consequently, the uncorrected date is probably a reasonably good estimator of the actual age of occupation of the Whelan Lake site.

Six radiocarbon assays of *Chione* sp. were conducted. The analysis was conducted by Sam Valastro of the University of Texas Radiocarbon Laboratory. Table 10 shows the uncorrected ^{14}C, ^{13}C corrected, and reservoir corrected dates for the Whelan Lake site shell samples. There was generally good internal agreement between the obtained shell dates. No samples were submitted from the disturbed top stratum, while most shell from the midden bottom was too fragmented for dating purposes. As Table 10 demonstrates, the uncorrected dates are clustered between 6,540 ± 90 B.P. and 7,300 ± 100 B.P., with an average of 7,150 ± 60 B.P. Of all samples submitted, the one from the most shallow provenience (0.20 m to 0.30 m depth) yielded the youngest date (6,540 ± 90 B.P.). The shell midden reaches a maximum depth of 1 m, but this maximum occurs in its center only and no test pits were placed here. In test pits, midden fill extended not further than to 0.70 m depth. The oldest date (7,300 ± 100 B.P.) comes from a sample collected from 0.30 to 0.40 m depth, while several samples from lower elevations yielded older dates. Apparently, temporal stratigraphy fluctuates somewhat across the midden. Based on absolute dating alone, there is no way of reliably determining the date of the initial occupation of the site. Nonetheless, we would estimate that the site was occupied from about 7,400 B.P. to 6,500 B.P., a 900 year period.

Table 10. Radiocarbon Determinations for Shell Samples from the Whelan Lake Site.

Test Pit #	Feature #	Level	Depth (cm)	Laboratory Number	Item #	Uncorrected C-14 Age (B.P.)	C-13 Adjusted Age (B.P.)	Reservoir Effect Corrected 415±35 Year Rate (B.P.)
		4	40-50	Tx-7406	117	7200±70	7640±80	6785±70
9		5	50-60	Tx-7407	190	7180±90	7610±90	6765±90
19		3	30-40	Tx-7408	173	7300±100	7730±100	6885±100
8		4	40-50	Tx-7409	241	7170±70	7620±80	6755±70
						7130±90	7560±90	6715±90
2	1	4	40-55	Tx-7410	266	7060±100	7500±100	6645±100
12	2	1	25-35	Tx-7511	279	6540±90	6980±100	6125±90

OBSIDIAN HYDRATION ANALYSIS

Paul D. Bouey, Thomas Origer and Rein Vanderpot

Five obsidian hydration samples were submitted to Paul Bouey from Far Western Anthropological research Group, Inc. for trace element evaluation. The analysis produced results consistent with most comparable work in southern California. Three pieces are assigned to the Coso Volcanic Field source area, and two to Obsidian Butte. Table 11 summarizes the findings from the analysis.

The trace element analysis of the obsidian specimens listed in Table 11 was conducted at the Department of Geology, University of California, Davis. This work was performed on a Kevex 0700 energy-dispersive X-ray fluorescence unit, using a rhodium (Rh) tube with a 0.05 mm rhodium filter at 30 kilovolts and 0.05 milliamps to analyze for rubidium (Rb), strontium (Sr), yttrium (Y), zirconium (Zr), niobium (Nb), lead (Pb), and thorium (Th). Only Rb, Sr, and Zr are reported here because those are the critical three elements used to differentiate obsidian sources in this region of the state. To acquire Fe/Mn ratios, a germanium (Ge) target run at 17 kilovolts and 1.7 milliamps was used. This unit has a Si (Li) detector and is used in conjunction with a Kevex 8000 multichannel analytical spectrometer.

Table 11. Trace Element Analysis of Obsidian Artifacts from the Whelan Lake Site.

SAMPLE	RB ppm	SR ppm	ZR ppm	SUM	RB %	SR %	ZR %	Fe/Mn	SOURCE
298	259.00	8.44	141.59	409.03	63.32	2.06	34.62	49.9	COSO
300	126.49	13.39	277.68	417.56	30.29	3.21	66.50	66.3	OB BUTTE
301	130.81	14.12	281.04	425.97	30.71	3.31	65.98	64.1	OB BUTTE
302	238.49	9.54	137.17	385.20	61.91	2.48	35.61	48.0	COSO
308	220.26	13.09	114.07	347.42	63.40	3.77	32.83	41.2	COSO

Normally in controlled geological studies, a rock is crushed into a powder and pressed into a pellet. That prepared sample has a homogenous distribution of constituent elements and is perfectly flat, providing the appropriate geometry for constant and systematic results. Archaeological materials cannot be treated in this manner, so their elemental distribution (glasses are assumed to possess homogenous elemental distributions) and imperfect geometry (i.e., their lack of a perfectly flat surface) must be compensated for through a form of "ratio" analysis.

In this procedure, samples were run for 200 live-seconds, and the resulting spectra were stripped of their backgrounds. Integrated densities were calculated for each of the elements in each sample, and those intensities were divided by the integral of the Compton peak. In conjunction with this work, geological rock standards with known quantities of constituent elements were similarly analyzed. The rock standard ratios were used to calculate a regression formula, and the sample ratios were used to derive the parts per million values for each specimen. Some of the samples in this collection are smaller than the accepted minimum size recognized for yielding reliable concentrations for the elements under study. Although those quantities may not be used to generate quantitative source characteristics, they still may be used to calculate ratios (i.e., Rb/Sr/Zr) in conjunction with Fe/Mn ratios which represent each of the various obsidian sources found in the region.

The quantitative computation used in this analysis is an acceptable method, although it is not a replacement for more detailed and accurate techniques (see Andermann and Kemp 1958; Nielsen 1979). Different machines and techniques also have been found to produce slightly varying numerical results, due, in part, to particle-size effects of the variously processed rock standards. Direct comparisons between laboratories are , therefore, problematic.

After the sources of the obsidian had been established, the five specimens were submitted for hydration band analysis. The analysis was completed by Thomas Origer at the Sonoma State University Obsidian Hydration Laboratory, an adjunct of the Anthropological Studies Center, Department of Anthropology.

Each specimen was examined in order to find two or more surfaces that would yield edges which should be perpendicular to the microslide when preparation of the thin section was completed. Two small parallel cuts were made at an appropriate location along the edge of each specimen with a 4-inch-diameter circular saw blade mounted on a lapidary trimsaw. The cuts resulted in the isolation of a small sample with a thickness of approximately one millimeter. Each sample was removed from its specimen and mounted with Lakeside Cement onto permanently etched petrographic microslides.

The thickness of the samples was reduced by manual grinding with a slurry of #500 silicon carbide abrasive on a glass plate. The grinding was completed in two steps. The first grinding was terminated when the sample's thickness was reduced by approximately 1/2, thus eliminating any micro-chips created by the saw blade during the cutting process. The slides were then reheated, which liquefied the Lakeside Cement, and the samples inverted. The newly exposed surfaces were then ground until the proper thickness was attained.

The correct thin section thickness was determined by the "touch" technique. A finger was rubbed across the slide, onto the sample, and the difference (sample thickness) was "felt." The second technique employed for arriving at proper thin section thickness is termed the "transparency" test. The microslide was held up to a strong light source and the translucency of the thin section observed. The sample was sufficiently reduced in thickness when the thin section readily allowed the passage of light.

A protective coverslip was affixed over the thin sections when all grinding was completed. The completed microslides are curated at the hydration lab under File No. 91-H1054.

The hydration bands were measured with a strainfree 40 power objective and Bausch and Lomb 12.5 power filar micrometer eyepiece on a Nikon petrographic microscope. Six measurements were taken at several locations along the edge of the thin section. The mean of the measurements was calculated and listed in Table 12 with other information. These hydration measurements have a range of ± 0.2 due to normal limitations of the equipment.

One specimen (no. 308) was marked by surfaces that appeared to be weathered. This condition was noted in Table 12 with a "w" under the "Remarks" column. Several places could be located that were free of weathered surfaces and at these the measurements were taken.

For the three Coso samples, absolute dates were extrapolated from Meighan's (1978) Coso obsidian index of 220 years/micron. Unfortunately, the resulting dates, 354 B.C., A.D. 196, and A.D. 494, although mutually fairly consistent, do not agree very well with the radiocarbon dates. Various factors may have been responsible for small, incompatible hydration measurements (compare Gallegos 1991:30). For this reason, the puzzling, young dates from these obsidian samples will be ignored. No chronometric readings can be established for the Obsidian Butte specimens, as no index has been compiled for this source. It is generally believed that the Obsidian Butte material is used later than Coso obsidian (Hughes and True 1985:333).

Table 12. Summary of Hydration Band Analysis of Obsidian from the Whelan Lake Site

CA-SDI-6010 October 1991

Lab#	Catalog#	Description	Provenience	Remarks	Measurements (microns)	Mean	Source
01	298	Debitage	TP 12/Level 1	none	7.2 7.2 7.2 7.3 7.4 7.4	7.3	Coso (x)
02	300	Debitage	TP 8/Level 3	none	7.3 7.3 7.4 7.5 7.5 7.6	7.4	Obs Butte (x)
03	301	Debitage	TP 2/Level 3	none	7.2 7.2 7.2 7.2 7.3 7.4	7.3	Obs Butte (x)
04	302	Debitage	TP 10/Level 3	none	10.5 10.6 10.7 10.7 10.7 10.7	10.7	Coso (x)
05	308	Debitage	TP 8/Level 3	w	8.1 8.1 8.1 8.2 8.4 8.5	8.2	Coso (x)

Lab Accession No.: 91-H0154 Technician: Thomas M. Origer

Conclusions

Radiocarbon dates obtained from six shell samples have established that the Whelan Lake site was occupied between 7,400 B.P. and 6,500 B.P. The dates from the individual samples are well-clustered and mutually consistent. Site composition and artifact content support the proposed age for the shell midden. The occurrence of a Pinto style projectile point in the midden fill is particularly befitting. Obsidian hydration band measurements correspond to dates that are perplexingly younger. Several of the obsidian specimens come from the same stratigraphic and spatial contexts as the shell samples, yet they yield dates ranging between 350 B.C. and A.D. 500. Thus, besides contradicting the radiocarbon dates, they also fall within the hiatus, from 1,000 B.C. to A.D. 500, observed in the archaeological sequence for this region. Two Tizon brown ware sherds, recovered from the midden surface, suggest that the site includes a later, San Luis Rey II component (circa A.D. 1500-1600). The small amount of ceramics recovered suggest that this component was unsubstantial, however. At this point, rather than explaining the young obsidian dates as denoting a later component, we believe that the obsidian specimens lack integrity, thereby electing the radiocarbon dates as the preferred time marker for the site.

CHAPTER 7
LITHIC ANALYSIS

Keith J. Vlastos

The lithic analysis for the SDI-6010 testing project involved the individual examination of each of the 718 lithic items recovered during excavation of the site. The goal of this analysis was to create a body of data relevant to the overall research questions addressed by the project. Questions specific to the lithic assemblage include: (1) what raw materials were exploited, (2) what technologies were used to exploit them, (3) were the raw materials exploited as part of a curated technology, or merely in an expedient fashion, and (4) are all raw materials exploited in the same fashion or are differences apparent in the exploitation of separate materials? Integral to answering these research questions is the identification of possible stratigraphic and horizontal variations in the occurrence of lithic materials. Such diversity may have temporal and functional implications for the site.

Two categories of lithic materials were recovered: flaked stone and ground stone. Each category was analyzed separately. Flaked stone artifacts are defined as the products and by-products of flint-knapping. Ground stone items are defined as the products of a ground stone technology. No by-products of ground stone technology have been identified.

The bulk of the lithic assemblage is made up of flaked stone (N=704) and includes bifacially worked items (N=4), cores (N=3), tools (N=19), and debitage (N=678). By far, debitage makes up the greatest portion of the flaked stone assemblage. The ground stone assemblage includes a single possible metate fragment, several manos (N=5), and a number of unidentifiable ground stone fragments (N=9).

FLAKED STONE

The objective of the flaked stone analysis was to identify and quantify the relative proportions of differing technological systems reflected by the lithic assemblage of the site. This approach allows the determination of the manner in which lithic materials were exploited by the occupants of SDI-6010, and in turn aids in defining the types and nature of activities performed there.

Methods

The flaked stone artifacts were divided into three categories: debitage, cores, and tools and tool fragments. The tools and tool fragments were further divided into four sub-categories: biface tools, modified flakes, choppers, and hammerstones. Many attributes, such as excavation unit, vertical level, and raw material were recorded for all lithic items in the assemblage. Attributes recorded for the 21 tools or tool fragments present in the assemblage include length, width, thickness, weight, and an identification of tool type. Measurements of length, width, and thickness were made to the nearest one-tenth millimeter using vernier calipers. These measurements were recorded only for complete tools. Weights were obtained using an Ohaus Dial-O-Gram and measured to the nearest tenth of a gram. The lithic assemblage from CA-SDI-6010 contains only three cores. Attributes recorded for each of these include amount of cortex present, core type, number of flake scars, and measurements of size and weight. Attributes recorded for debitage include raw material type, amount of cortex present, platform type, size, technology, and debitage type. Discussions of these attributes and how they were recorded are presented with the results of the analysis.

Results

Raw Materials

An important part of any lithic analysis is the identification of the raw materials used by the flint-knappers. Eight classes of raw material were identified in the assemblage from the Whelan Lake site. Table 13 illustrates the distribution of raw materials within the debitage from the flaked stone assemblage.

Table 13. General Distribution of Raw Materials.

CCS	PDL Chert	Quartz/ Quartz Crystal	Basalt	Granite	Quartzite	Fine-Grained Basalt	Obsidian	Total #
40 (5.9%)	28 (4.1%)	64 (9.4%)	262 (38.6%)	3 (0.5%)	239 (35.3%)	37 (5.5%)	5 (0.7%)	678

KEY: CCS=Cryptocrystalline Silicate; PDL chert=Piedra del Lumbre chert.

Two material types comprise the bulk of the assemblage, basalt and quartzite. Basalt is an extrusive igneous rock consisting of plagioclastic crystals not visible to the naked eye. It is characterized by its dark color and dense consistency. Numerous cobbles of basalt occur along the beach 10 km to 15 km west of SDI-6010. Though the beach may have been somewhat further away during prehistoric times, the basalt exploited at the site was most likely derived from a similar coastal environment. A basalt of slightly finer-grained consistency was present in smaller quantities within the flaked stone assemblage. This fine-grained basalt was most likely obtained from from either sparse coastal sources or unknown inland sources.

Quartzite is commonly formed by the metamorphism of sandstone. It is composed of quartz and exhibits poorly developed conchoidal fractures. Quartzite cobbles appear along the coast less frequently than basalt, but are also known to occur in the canyons of Camp Pendleton 8 km to 20 km north of SDI-6010 (Ezell 1977). Much of the quartzite exploited at SDI-6010 was probably obtained from the coast where ease of procurement made it an attractive material.

Of the remaining materials, quartz and quartz-crystal appear most frequently. Quartz is a mineral formed from silicon dioxide (SiO_2). Usually white in color, it also occurs in yellow, pink, and pale brown to black. In its crystalline form, it is clear and colorless. Quartz exhibits a greasy luster and a weak to good conchoidal fracture. Much of the quartz recovered at the Whelan Lake site was shatter of poor quality for flint-knapping. Gibson and Sandler (1990:E.5) report outcrops of milky white quartz and quartz-crystal within 5 km of SDI-6010, and Ezell (1977:50) reports an abundance of the material in the Camp Pendleton area.

In the site's chipped stone assemblage, granite is represented only by shattered rock and a single hammerstone. No doubt this reflects its poor quality for use in flint-knapping. In contrast, the ground stone assemblage from the site is made up entirely of granite cobbles and cobble fragments. The acquisition of granite is similar to that of basalt and quartzite. Appearing in beach cobble deposits at the coast, this coarse-grained intrusive igneous rock was easy to obtain and transport.

Chert, chalcedony, agate, and jasper are all sedimentary rocks formed by the compaction and partial recrystallization of siliceous skeletons, the precipitation of silicas from sea water, or the partial replacement of limestone with silica. These materials exhibit a highly developed conchoidal fracture and a waxy to vitreous luster. The well-developed conchoidal fracture and smooth texture of these materials makes them well-suited to flint-knapping techniques. For the purposes of this analysis, these materials have been classified together as cryptocrystalline silicates (CCS). Outcrops of CCS are found in canyons of Camp Pendleton approximately 10 km to 20 km north of SDI-6010.

A specific form of CCS outcrops at Camp Pendleton in Piedra del Lumbre (PDL) Canyon approximately 18 km north of Whelan Lake. A large prehistoric quarry site is located in the canyon (Gibson and Sandler 1990:E.5). PDL chert ranges in color from rusty brown to buff. Its texture is very fine with numerous quartz-crystal inclusions which give it a sparkling appearance (hence the name of the canyon). PDL chert accounts for a relatively small proportion (4.1%) of the debitage recovered at SDI-6010.

Five flakes of obsidian were recovered during testing at SDI-6010. Obsidian is a non-crystalline, amorphous rock, formed by the flow of silica-rich magma that solidifies before minerals can crystallize. Most commonly it is black with smoky translucent or transparent edges. It is a hard material (6 or 7 on the Moh's scale) which exhibits strong conchoidal fractures due to the homogeneity of its structure. Thus, it is particularly well-suited to flint-knapping techniques. Obsidian does not occur naturally within the San Luis Rey River drainage and had to be acquired from some distance.

Chapter 6 presented data from the X-ray fluorescence analysis of the obsidian samples recovered at SDI-6010. Three of the samples match trace element profiles of obsidian from the Coso volcanic field more than 400 km north of the site. The two remaining samples match profiles from Obsidian Butte in extreme southern California southeast of the Salton Sea. It has been suggested that Coso obsidian was traded along a coastal route into northwestern San Diego County (Hughes and True 1985:333). Obsidian Butte material was traded throughout the interior and is the closest known source to SDI-6010 (Gibson and Sandler 1990:E.4).

The presence and proportion of each material in the assemblage from SDI-6010 provides insight into the activities of the occupants of the shell midden. The large proportions of basalt and quartzite speak not only of the ease of procurement of these materials, but also of their relative value to other materials. The low percentages of CCS, PDL chert, and obsidian debitage present in the lithic assemblage indicate that smaller amounts of these materials were exploited. It also suggests that because these materials were more difficult to obtain, they were exploited much more economically. Finally, the presence of obsidian at SDI-6010 clearly demonstrates that the prehistoric inhabitants of the area were tied into trade networks which extended throughout southern California.

Cortex

Cortex has been defined as the natural surface or rind on flint-like materials (Crabtree 1972:30). The primary reduction of raw materials often involves the removal of cortex and other poor quality material from a core prior to tool manufacture or transport. Therefore, the amount of cortex present on various raw materials can be indicative of the relative distance from the geologic source an item has been transported by the aboriginal procurement system. It may also indicate whether the material is being exploited as part of an efficient and curated technology or in an expedient manner. The amount of cortex present on the dorsal surface of each item of debitage was recorded as either 0%, less than 50%, 50-99% or 100%

As shown in Table 14, the majority of each lithic material from SDI-6010 contains little or no cortex. The low proportion of cortex within the assemblage suggests that the materials utilized at the Whelan Lake site have, for the most part, undergone the process of primary reduction elsewhere. The absence of cortex on the obsidian flakes is not surprising considering the exotic nature of the material. In contrast, the lack of cortex on the fine grained basalt suggests that this material may have been procured elsewhere, even though the material is available close by. Overall, the low percentage of cortex in the assemblage indicates the practice of a well-thought-out, curated technology involving the importation of previously reduced materials, and possibly existing tools, to the site for use.

Table 14. Amount of Cortex Present in Debitage.

% Cortex	CCS	PDL Chert	Quartz/ Quartz Crystal	Basalt	Quartzite	Fine-Grained Basalt	Obsidian
0	38 (95.0%)	26 (93.0%)	57 (89.0%)	235 (89.7%)	206 (86.2%)	37 (100.0%)	5 (100%)
<50	2 (5.0%)	1 (3.5%)	4 (6.3%)	15 (5.7%)	14 (5.9%)		
50-99		1 (3.5%)	3 (4.7%)	12 (4.6%)	18 (7.5%)		
100					1 (0.4%)		
Total #	40	28	64	262	239	37	5

Flake Size

The examination of flake size may yield information concerning both the technologies utilized and the distance of a site from a procurement location. Simply stated, if debitage, as a by-product of tool manufacture, is truly "waste material" it is unlikely it will be transported over long distances by pedestrian hunter-gatherers (Towner 1991:216). Likewise, the size and amount of debitage of any given material will reflect how efficiently that material is being exploited.

The debitage in the assemblage from SDI-6010 was broken down into three types: flakes, shatter, and fire-cracked debris. Flakes include any lithic artifact with a discernible interior surface which does not exhibit signs of further cultural modification. Both complete and incomplete flakes were included in this category. Any debris from the reduction process which does not contain a bulb of percussion, platform or a clearly discernible interior surface was classified as shatter. Debris resulting from exposure to fire also contain no bulb, platform or interior surface, but exhibit signs of heating such as spalls, cracks or discoloration (Harry 1990:123). Items of this type were classified as fire-cracked debris. Only two pieces of fire-cracked debris were collected at SDI-6010: one of quartzite and one of granite.

Before examining information concerning flake size, it may be helpful to look at the amount of shatter in the assemblage. As stated above, lithic shatter is debris of the reduction process. Whereas flakes reflect the application of specific, intentional technologies, shatter is often the result of a much more expedient form of reduction, involving the removal of unwanted or poor quality material from a core or blank prior to tool manufacture or transport. Therefore, the proportion of shatter present within any given material type should offer information concerning the amount of waste produced during the reduction and/or exploitation of that material. Table 15 presents the proportion of flakes and shatter for each material type.

Table 15. Proportion of Flakes in Debitage.

Debitage Type	CCS	PDL Chert	Quartz/ Quartz Crystal	Granite	Basalt	Quartzite	Fine-Grained Basalt	Obsidian
Flake	38 (95.0%)	26 (92.9%)	12 (18.8%)		213 (81.3%)	182 (76.2%)	32 (86.5%)	5 (100.0%)
Other	2 (5.0%)	2 (7.1%)	52 (81.2%)	3 (100.0%)	49 (18.7%)	57 (23.8%)	5 (13.5%)	
Total #	40	28	64	3	262	239	37	5

Of the quartz recovered at SDI-6010, 81.2% is shatter of poor quality material, indicating primary reduction of this material was being performed on site. This suggests a source in close proximity to the site, possibly even closer than the outcrop reported by Gibson and Sandler (1990:E.5). The occurrence of relatively small proportions of shatter within the remaining material types indicates that these materials underwent initial reduction before being brought to Whelan Lake.

Of the debitage from SDI-6010, 75.1% is composed of flakes of varying sizes. The maximum dimension of each flake was measured using a template of concentric circles placed 2 cm apart. Flakes less than 2 cm, between 2 cm and 4 cm, 4 cm and 6 cm, and 6 cm to 8 cm in size were present. No flakes larger than 8 cm in any dimension were recovered during testing at SDI-6010. Table 16 shows the distribution of different sized flakes for each raw material type.

The presence of large flakes in the debitage for any given raw material type may indicate a less economic exploitation of that material. Table 16 reveals that the obsidian from SDI-6010 has been exploited in a very economical fashion, producing only a small number of flakes less than 2 cm in size. This underlines the exotic nature of this material and its relative value in comparison to the other raw material types. Of the remaining materials, CCS, PDL chert and fine-grained basalt also appear to have been exploited economically. CCS and PDL chert have known sources at some distance from the site, and while the source of fine-grained basalt remains in some doubt, this information suggests its curation and exploitation was similar to these other materials. Quartz, basalt and quartzite appear to have been the least valued of the materials exploited at SDI-6010. Yet, even for these material types only a small proportion of flakes larger than 2 cm in size appear. Clearly, a very economic and conscientious exploitation of all lithic materials was being practiced at this site.

Table 16. Flake Size.

Flake Size (cm)	CCS	PDL Chert	Quartz/ Quartz Crystal	Basalt	Quartzite	Fine-Grained Basalt	Obsidian
0-2	35 (92.1%)	23 (88.5%)	10 (83.4%)	171 (80.3%)	145 (79.7%)	30 (93.8%)	5 (100.0%)
2-4	3 (7.9%)	2 (7.7%)	1 (8.3%)	40 (18.8%)	32 (17.6%)	1 (3.1%)	
4-6		1 (3.8%)		2 (0.9%)	3 (1.6%)	1 (3.1%)	
6-8			1 (8.3%)		2 (1.1%)		
Total #	38	26	12	213	182	32	5

Technologies

Crabtree (1972:94) defines technology as the sum and organization of techniques used to manufacture a specific item. Tools and debitage from SDI-6010 indicate the presence of at least three different technological systems: a flake-core technology, a biface technology, and a bi-polar technology. Due to the small number of tools recovered, the identification of technologies was based primarily on debitage data.

Information recorded for each flake in the assemblage included an assignment into one of seven technological categories:

(1) Bi-polar Reduction. Bi-polar flakes exhibit evidence that force has been applied to both ends of the flake. This includes crushing on both ends, shattered or pointed platforms, pronounced ripple marks, no definite positive bulb of force, and generally parallel sided form. Bi-polar flakes are produced during a bi-polar reduction process involving resting a core upon an anvil then striking it to produce a flake. Other reduction techniques may also occasionally produce flakes which exhibit bipolar characteristics so care is needed when classifying items as bi-polar.

(2) Primary Reduction. Flakes which exhibit a definite positive bulb of percussion and have all or nearly all of their dorsal surface covered by cortex were classified as primary reduction flakes. These flakes are produced during the initial reduction of raw materials and are indicative of a flake-core technology.

(3) Secondary Reduction. The secondary reduction of raw materials typically produces flakes with definite positive bulbs of percussion and the majority of their dorsal surface covered with cortex. Flakes exhibiting these characteristics were classified as secondary reduction flakes and are also considered indicative of flake-core technology.

(4) Bifacial Thinning. Bifacial thinning flakes are flakes which result from either percussion or pressure flaking during operations designed to thin a core or manufacture a biface. These flakes display most or all of the following characteristics: a thin, curved, longitudinal cross-section; small lateral and distal edge angles; multiple flake scars on the dorsal surfaces originating from several directions; a multifaceted striking platform; little or no presence of cortex; an expanding shape in plan view; and a diffuse bulb of force.

(5) Pressure Flakes. Bifacial thinning flakes measuring less than 0.5 cm in length from the bulb of applied force to successful termination on the distal edge were classified as pressure flakes. These flakes are a particular type of bifacial thinning flake produced

during the retouching of tool edges during tool manufacture or the resharpening of existing tool edges.

(6) General Flakes. Flakes which do not exhibit any of the attributes of the above flake types were classified as general flakes.

(7) Indeterminate. Any flakes which could not be assigned to a particular category were considered technologically undiagnostic and classified as indeterminate (Harry 1990:123-124). Only diagnostic flakes appear in the following tables. Indeterminate flakes (N=184) were not considered.

Table 17 illustrates several important trends. Most apparent of these is the predominance of biface technology (bifacial thinning flakes and pressure flakes) in all material classes. A closer examination of this trend reveals that not only has 100 percent of the obsidian recovered been bifacially exploited, but 96.8 percent of the CCS diagnostic flakes are by-products of bifacial technology. Four materials show some amounts of primary reduction: quartz/quartz-crystal, PDL chert, quartzite, and basalt. In fact, 30 percent of the quartz/quartz-crystal diagnostic flakes are either primary or secondary reduction flakes, but due to the small sample size (N=10) this trend may not be representative of the entire assemblage. Interestingly, only a small proportion (8.6% and 14.6%, respectively) of basalt and quartzite appears to have undergone primary or secondary reduction on site. Bi-polar technology accounts for a mere 2.6 percent of all diagnostic flakes recovered and is seen predominantly within the harder, less brittle material classes.

Cores

As mentioned above, three cores were recovered during testing at the Whelan Lake site. A core is defined as a piece of lithic material which has had flakes removed from it. It is a nucleus exhibiting negative flake scars (Crabtree 1972:30). For the purpose of this analysis however, formal tools exhibiting pressure flaking or retouch have been excluded from this definition.

Table 17. Distribution of Diagnostic Flakes in Each Raw Material.

Technology	CCS	PDL Chert	Quartz/ Quartz Crystal	Basalt	Quartzite	Fine-Grained Basalt	Obsidian
Primary Reduction		1 (4.8%)	1 (10.0%)	2 (1.4%)	7 (5.7%)		
Secondary Reduction			2 (20.0%)	10 (7.2%)	11 (8.9%)		
Bi-polar		1 (4.8%)		3 (2.2%)	3 (2.5%)	2 (9.1%)	
Bifacial Thinning	23 (74.2%)	13 (61.9%)	5 (50.0%)	81 (58.3%)	73 (59.4%)	10 (45.5%)	5 (100.0%)
Pressure Flakes	7 (22.6%)	1 (4.7%)	1 (10.0%)	20 (14.4%)	11 (8.9%)	4 (18.2%)	
General Flake	1 (3.2%)	5 (23.8%)	1 (10.0%)	23 (16.5%)	18 (14.6%)	6 (27.2%)	
Total #	31	21	10	139	123	22	5

Two core types were identified: uni-directional and bifacial. Uni-directional cores exhibit flake removals from a single direction and, most often, from a single platform. Bifacial cores, on the other hand, can be identified by flake removals from opposite margins and opposite faces (Crabtree 1972:16).

Two of the three cores recovered at SDI-6010 are unifacial, one of quartzite and smaller core of basalt. These represent a more expedient technological approach than does the single bifacial core of quartzite. Because bifacial cores yield flakes with more usable edges than other types of cores, they are often associated with curated technologies (Goodyear 1979).

While it is difficult to draw conclusions from such a small sample of cores, it is worth noting that the presence of the more expedient uni-facial cores is not well reflected in the debitage. Also, the absence of cores of any materials other than quartzite and basalt argues for the presence of a curated and conscientious technology involving these other materials. It appears that finished and nearly finished bifaces of these materials were brought to the site and utilized.

All three of the cores are cobbles or cobble fragments which have been exploited for flake removal. The bifacial core exhibits signs of battering on the end opposite flake removal, indicating its use as a hammerstone.

Tools

Of the 23 tools and tool fragments recovered during testing at Whelan Lake, seven were collected from backhoe trench back-dirt and are unprovenienced. One was surface collected in the area of Test Pit 8. The tools recovered include edge modified flakes (N=7), hammerstones (N=10), choppers (N=2), and four biface tool fragments (Table 18).

Edge Modified Flakes

Six scrapers were recovered at SDI-6010. These tools were manufactured from large, primary reduction flakes. The end scrapers are beveled tools made on flakes with a working edge on one or both convex ends (Figure 26a). The side scraper is an irregular flake with one curved border modified by unifacial flaking (Figure 26b). These scrapers were most likely manufactured on the site in an expedient fashion during the reduction of basalt and quartzite cobbles.

Pics are bifacial stoneworking tools (Figure 27), first defined in San Diego County in the lithic assemblage from SDI-5130 (Gibson and Sandler 1990:E.27). These tools exhibit heavy abrasion, battering, crushing, and grinding scars on one or more edges. They appear to have been used primarily in the quarrying of granite slabs and in pecking, shaping, and grinding metates, manos, and other ground stone implements. A single pic of fine-grained basalt was recovered at SDI-6010. It was probably utilized in shaping ground stone implements at the site. Since this pic was manufactured from a flake struck from a cobble, it can be assumed at least some of the fine-grained basalt used at SDI-6010 was originally procured from secondary or tertiary geologic sources along the coast.

No flakes exhibiting micro-flaking and/or polish on one or more surfaces or other modification due only to wear were identified in the debitage from SDI-6010. Flakes exhibiting these characteristics are considered utilized flakes. The small size of the majority of the flakes in the assemblage make them unsuitable for utilization. The examination of flakes for use-wear was performed only with the naked eye. It is possible that under microscopic examination, utilized flakes may be identified within the assemblage.

Table 18. Tool Distribution.

Test Unit	Level (cm)	Raw Material	Length (cm)	Width (cm)	Thickness (cm)	Weight (gms)	Comments
2	20-30	PDL Chert	2.61	1.43	0.57	2.53	bifacial drill frag.
4	30-40	Quartzite	8.13	7.27	3.01	76.51	end scraper
4	30-40	Basalt	9.73	8.44	4.56	388.1	unmodified hammerstone
5	30-40	Basalt	6.58	5.27	2.15	71.23	side scraper
5	70-80	Basalt				75.32	large biface tip
6	10-20	Basalt	7.57	6.83	2.86	233.13	unifacial hammerstone
6	40-50	Quartzite	6.61	4.83	3.22	106.19	bifacial hammerstone
7	40-50	Basalt	8.47	7.64	4.32	326.44	unmodified hammerstone
8	10-20	Quartzite	8.35	5.97	1.82	89.72	end scraper
8	10-20	Quartzite				9.16	biface fragment
8	30-40	Basalt	7.91	5.33	3.67	205.87	unmodified hammerstone
10	40-50	Basalt				10.31	projectile point base
11	10-20	Basalt	5.13	4.71	2.16	82.02	end scraper
12	0-10	Basalt	4.04	3.73	3.83	71.69	unmodified hammerstone
12	0-30	Basalt				65.50	small hammer or pecking stone
P.P.		Basalt				32.01	small hammer or pecking stone from area of Testpit 8
N.P.		Fine Grained Basalt	6.92	6.74	3.17	183.73	pic from Trench A
N.P.		Quartzite	12.72	9.52	4.84	602.85	chopper from Trench A
N.P.		Basalt	13.47	10.61	3.12	547.76	chopper from Trench C
N.P.		Quartzite	8.81	7.81	3.63	363.7	unmodified hammerstone from Trench C
N.P.		Granite	10.27	6.52	3.81	296.28	bifacial hammerstone from Trench E
N.P.		Basalt	7.82	5.07	2.62	96.49	end scraper,unifacialy retouched, from Trench G
N.P.		Quartzite	6.53	5.41	2.24	116.45	end scraper from Trench H

KEY: N.P.=Non-Provenienced;P.P.=Point Provenienced.

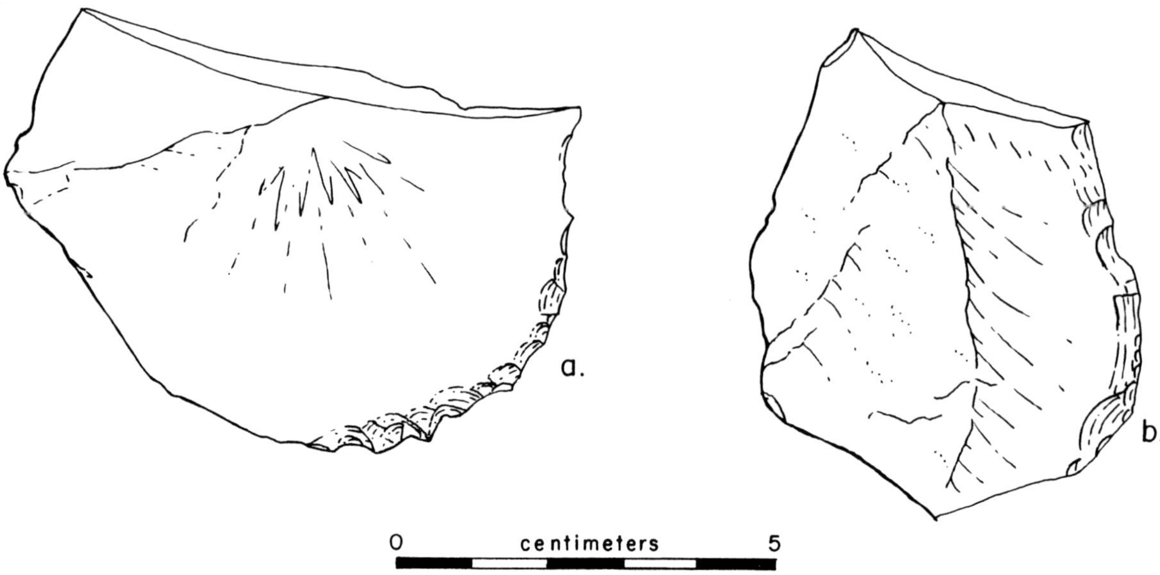

Figure 26. End Scraper (a) and Side Scraper (b).

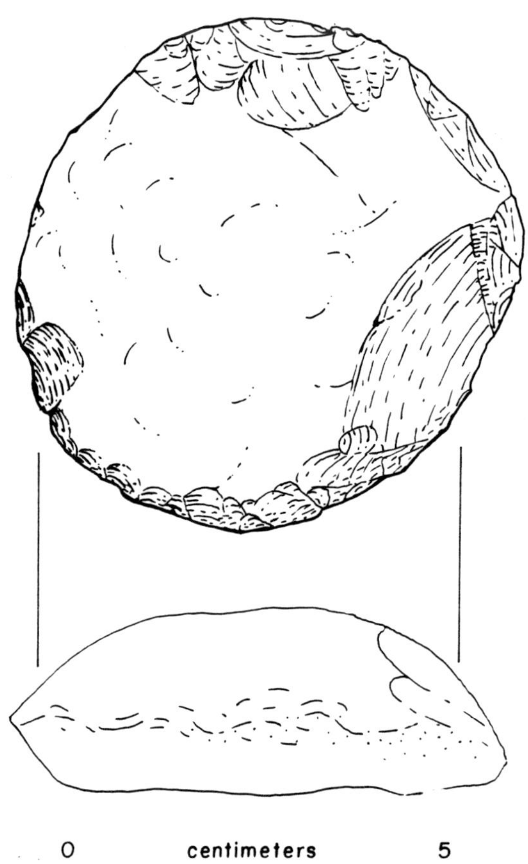

Figure 27. Pic from Trench A.

Hammerstones

Hammerstones exhibit pecking and battering at one or more locations on one or more ends. Their most frequent use was in the manufacture of flaked stone implements, but hammerstones were used for a wide range of functions that require precise blows from a hard, hand-held object. Of the ten hammerstones in the assemblage, five are otherwise unmodified, fist-sized cobbles; two show modification by bifacial flaking to the working edge; one displays unifacial flaking on its working edge; and two are small basalt pebble fragments exhibiting light pecking and abrasion on one end.

Basalt is the most common material among the hammerstones. The presence of a granite hammerstone, however, suggests proper sized stones of any material on hand may have been utilized for hammerstones.

Choppers

Two choppers were recovered from back dirt alongside backhoe trenches. These are cobbles with unifacial flake scars. They are heavy instruments that are percussion flaked and trimmed to form a jagged cutting edge along one or more sides. Both choppers from SDI-6010 have one unmodified surface.

Biface Tools

Biface tools are defined simply as any lithic item which exhibits pressure flake scars on two faces (Crabtree 1972:38). The SDI-6010 assemblage contains only four biface items, none of which are completely intact.

A drill fragment of PDL chert was recovered from Test Pit 2 (Figure 28a). The tip of the drill is missing and it appears a bi-polar flake was removed from the base after fracture, illustrating the thoroughness with which materials were utilized and recycled at the site.

Of the remaining biface items, two are of basalt (Figure 28b) and one of quartzite (Figure 28c). It appears likely that these items were fractured prior to deposition. The projectile point base (Figure 28b) and large biface tip (Figure 28d) were both recovered from levels well below any zone of plow disturbance that may have existed. The base contains a quartz inclusion along its fracture, suggesting it may have fractured during manufacture. Finally, the two parts of the smaller biface fragment (Figure 28c) were recovered separately in Test Pit 8.

The projectile point base is stylistically similar to what other investigators have termed Pinto points in the Mojave Desert (Harrington 1957). This type has a wide distribution ranging from the type site in the Mojave Desert throughout the Great Basin and into much of the greater Southwest. Pinto points are characterized by a certain thickness and crudeness, percussion-chipping, a frequent use of serrated edges, and in most cases by pronounced basal notches often producing protruding "ears" (Harrington 1957:49). Dating of the Pinto Basin tradition has proven problematic, yet the consensus is that the tradition emerged sometime during the Late Archaic no later than 3,000 B.C. lasting possibly into the Middle Pacific Period (500 B.C. to A.D. 500) (Chartkoff and Chartkoff 1984:124).

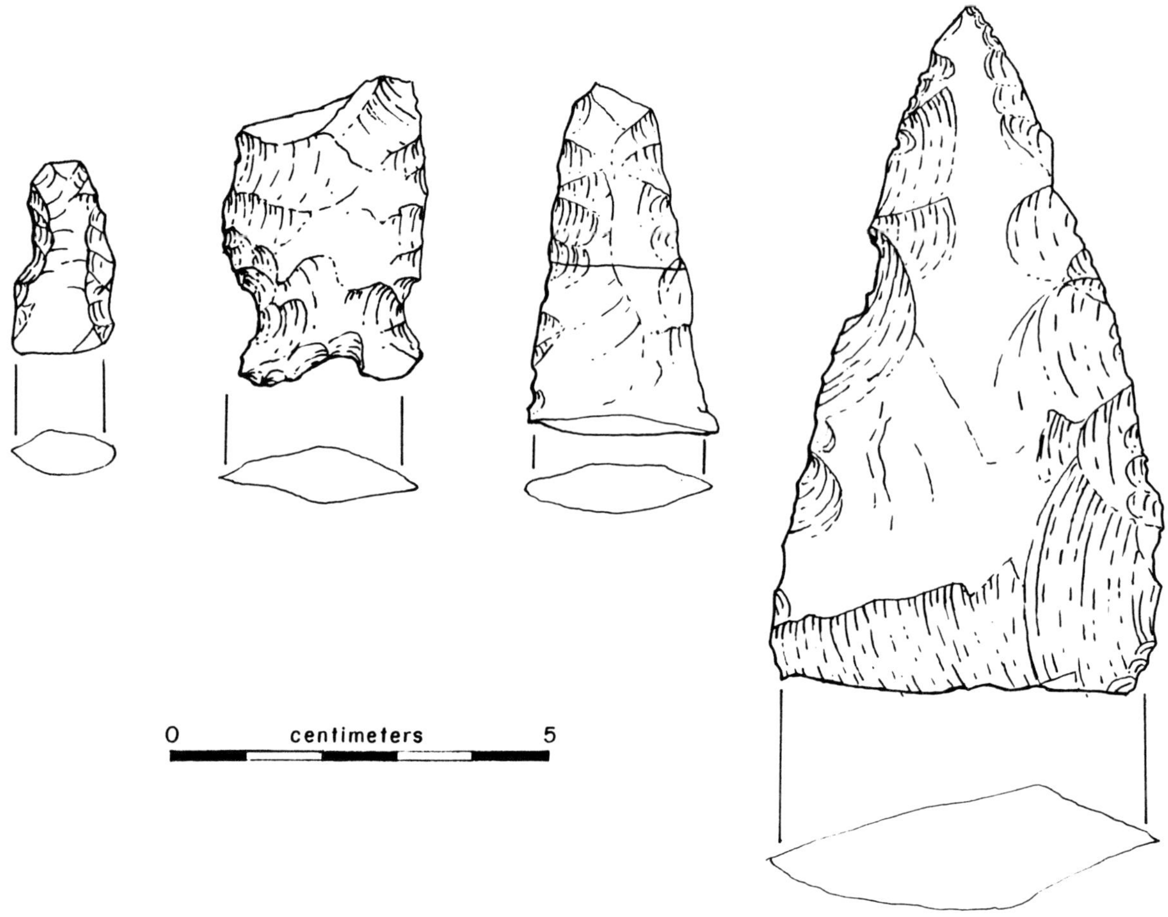

Figure 28. Biface tool fragments: (a) drill tip, (b) projectile point base, (c) biface fragment, (d) biface tip.

Obviously, it is unsound to type an assemblage using only a single diagnostic item. The lack of more and better diagnostic items within the chipped stone assemblage from SDI-6010 make the relative dating of the assemblage impossible. Local inhabitants of the Whelan Lake area have stated that projectile points have been collected from the site, though this could not be confirmed. The absence of these items is notable in light of the amount of bifacial debitage within the assemblage. Apparently, many of the tools manufactured at the site were curated, utilized, or discarded elsewhere.

Flaked Stone Distribution

The horizontal and vertical distribution of flaked stone artifacts at the site may help in identifying activity areas and occupation zones. Table 19 presents the vertical distribution of raw materials at SDI-6010. Materials from Test Pits 3 and 12 have been excluded from the following tables. Test Pit 3 lies outside the site boundaries, and Test Pit 12 was not excavated in ten centimeter levels.

Table 19. Vertical Distribution of Raw Materials.

Level (cm)	CCS	PDL Chert	Quartz/ Quartz Crystal	Basalt	Quartzite	Fine-Grained Basalt	Obsidian
0-10	7 (18.4%)	4 (15.4%)	5 (8.6%)	41 (17.0%)	42 (19.3%)	1 (2.7%)	
10-20	11 (28.9%)	6 (23.1%)	14 (24.2%)	55 (22.8%)	54 (24.8%)	8 (21.6%)	
20-30	11 (28.9%)	9 (34.6%)	22 (37.9%)	75 (31.1%)	64 (29.3%)	16 (43.3%)	4 (100.0%)
30-40	2 (5.3%)	3 (11.5%)	4 (6.9%)	31 (12.9%)	13 (6.0%)	8 (21.6%)	
40-50	5 (13.3%)	2 (7.7%)	8 (13.8%)	25 (10.4%)	33 (15.1%)	3 (8.1%)	
50-60	1 (2.6%)		5 (8.6%)	5 (2.1%)	9 (4.1%)		
60-70	1 (2.6%)	2 (7.7%)		8 (3.3%)	3 (1.4%)	1 (2.7%)	
70-80				1 (0.4%)			
Total#	38	26	58	241	218	37	4

Table excludes items from test units 3 and 12.

Table 19 indicates that the main occupation zone at SDI-6010 occurs from 0 cm to 30 cm below surface. In fact, the top three ten centimeter levels account for over 60 percent of the chipped stone artifacts recovered. The higher density of lithics within the upper levels of excavation may be due in part to the deflation of the midden over time. However, the sharp decrease in lithic density found at 30 cm below surface and again at 50 cm below surface, may indicate two possible occupations. Table 19 also shows that the highest proportion of each raw material occurs between 20 cm and 30 cm below surface, indicating an increase in activity at the site.

The flaked stone distribution is similar to that observed in the shell remains (see Chapter 8). Peak intensity of site use corresponds to levels 3 and 5. This correspondence is site wide, indicating that all areas of the site were utilized simultaneously. The bimodal distribution of artifacts suggests that the occupation of the site was episodic, with periods of intense use fluctuating with those of ephemeral visits.

The proportion of raw materials present within each level is presented in Table 20. This table once again illustrates a reliance upon basalt and quartzite throughout the occupation at SDI-6010. It also demonstrates that other raw materials were consistently exploited as well. Level 3 reflects the most diverse exploitation of raw materials noted for all materials. Thirty percent of the lithic raw materials occurring in this level are other than basalt or quartzite. Overall, Tables 19 and 20 show that there is little temporal variation in the types of materials exploited at SDI-6010.

Table 21 shows the proportion of each technology present in the debitage from each level. The dominance of bifacial technology appears consistent throughout the occupation of SDI-6010 with the diversity of technologies apparently increasing in the upper levels. Not surprisingly, the greatest technological diversity appears between 0 cm and 30 cm below surface. Rather than reflecting a true change in the technological exploitation of raw materials , the diversity seen in the upper levels of the cultural stratum is most likely due to an increased usage of the site. Any increase in site activity would require the exploitation of a greater quantity of lithic material. This increase in quantity is reflected in the data of Table 19. The increased use of lithic materials may in turn require a greater exploitation of readily available materials that required at least some degree of primary and secondary reduction on site. The data in Tables 21 and 17 support this assertion. Lithic technologies at SDI-6010 seem to display very little temporal variation. The production and exploitation of bifacial tools appears to have been of primary importance throughout the occupation of the site.

Table 20. Vertical Proportions of Raw Materials.

Level (cm)	CCS	PDL Chert	Quartz/ Quartzite	Basalt	Quartzite	Fine Grained Basalt	Obsidian	Total #
0-10	6.9%	4.0%	4.9%	40.6%	41.6%	1.0%		101
10-20	7.4%	4.1%	9.4%	37.2%	36.5%	5.4%		148
20 30	5.5%	4.5%	10.9%	37.3%	31.8%	8.0%	2.0%	201
30-40	3.3%	4.9%	6.6%	50.8%	21.3%	13.1%		61
40-50	6.5%	2.6%	10.4%	32.5%	42.8%	3.9%		77
50-60	4.8%		23.8%	23.8%	42.8%			21
60-70	6.7%	13.3%		53.3%	20.0%	6.7%		15
70-80				100.0%				1

Table excludes items from test units 3 and 12.

Table 21. Vertical Distribution of Diagnostic Flakes.

Level (cm)	Primary Reduction	Secondary Reduction	Bi-polar	Bifacial Thinning	Pressure Flakes	General Flake	Total #
0-10	1 (1.8%)	6 (10.9%)		39 (70.9%)	9 (16.2%)		55
10-20	2 (2.2%)	6 (6.5%)		44 (47.8%)	15 (16.3%)	25 (27.2%)	92
20-30	5 (5.9%)	2 (2.4%)	5 (5.9%)	40 (47.0%)	16 (18.8%)	17 (20.0%)	85
30-40				23 (82.1%)		5 (17.9%)	28
40-50		7 (17.1%)	1 (2.4%)	30 (73.2%)	3 (7.3%)		41
50-60	1 (8.3%)		1 (8.3%)	8 (66.7%)		2 (16.7%)	12
60-70				5 (55.6%)		4 (44.4%)	9
70-80							0

Table excludes items from test units 3 and 12.

Figure 29 is a contour map based upon the quantity of debitage recovered from each of the test units within the site boundaries. The total quantity of lithic material from each test unit was used to establish a lithic density per cubic meter. For Test Pits 8 and 2, both started as 2 m x 2 meter units. Then boundaries containing units of similar density were drawn.

The highest concentration of lithic materials is in the southeast quadrant of the site and concentrations remain relatively high throughout the area containing the four hearth features. Test Pits 1 and 7 on the west side of the site, and 4 and 5 on the north contain the lowest concentrations of debitage. This may indicate that these units were placed in areas peripheral to the site's main activity area.

Analysis of the debitage from SDI-6010 reveals that lithic production at the site primarily involved the manufacture of biface tools from previously reduced materials. It is likely that most evidence of this process occurs in the site's activity areas, if they exist.

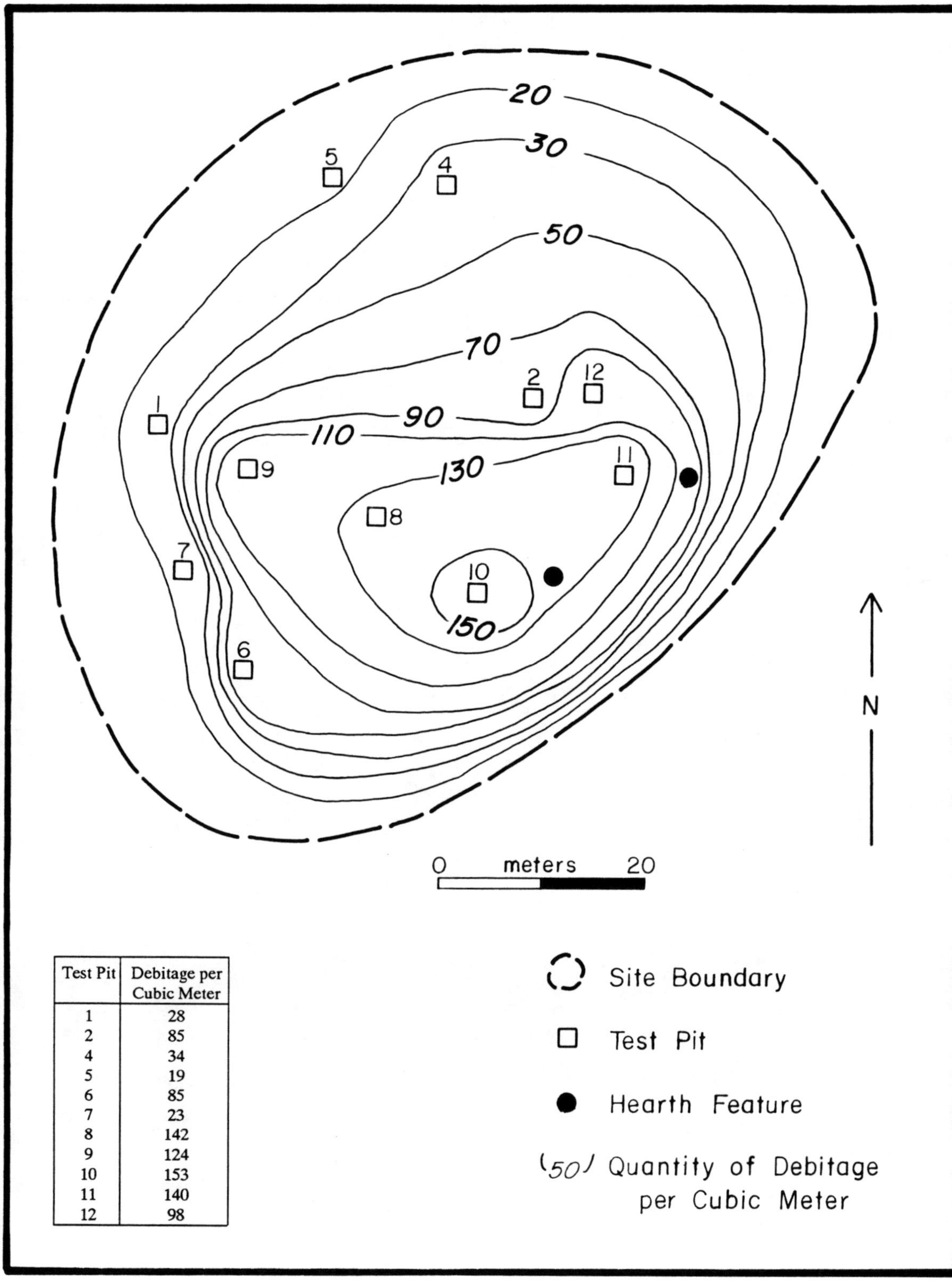

Test Pit	Debitage per Cubic Meter
1	28
2	85
4	34
5	19
6	85
7	23
8	142
9	124
10	153
11	140
12	98

Figure 29. Lithic density.

Basalt and quartzite formed a key component in the site's lithic manufacture and utilization. Figure 30 is a distribution map based upon the ratio of other lithic materials to basalt and quartzite recovered from each test unit.

The western edge of the site, in the vicinity of Test Pits 1, 4, and 7, exhibit a higher proportion of other materials, whereas the exploitation of basalt and quartzite is most pronounced in the southeastern portion of the site where the lowest ratios occur. This indicates a more concentrated utilization of basalt and quartzite in the southeastern portion of the site.

Test Pit	Basalt and Quartzite	Other Materials	Ratio
1	6	7	1.17
2	59	25	0.42
4	11	6	0.54
5	9	4	0.44
6	43	13	0.30
7	10	6	0.60
8	165	63	0.38
9	68	19	0.28
10	58	11	0.19
11	30	12	0.40
12	31	8	0.26

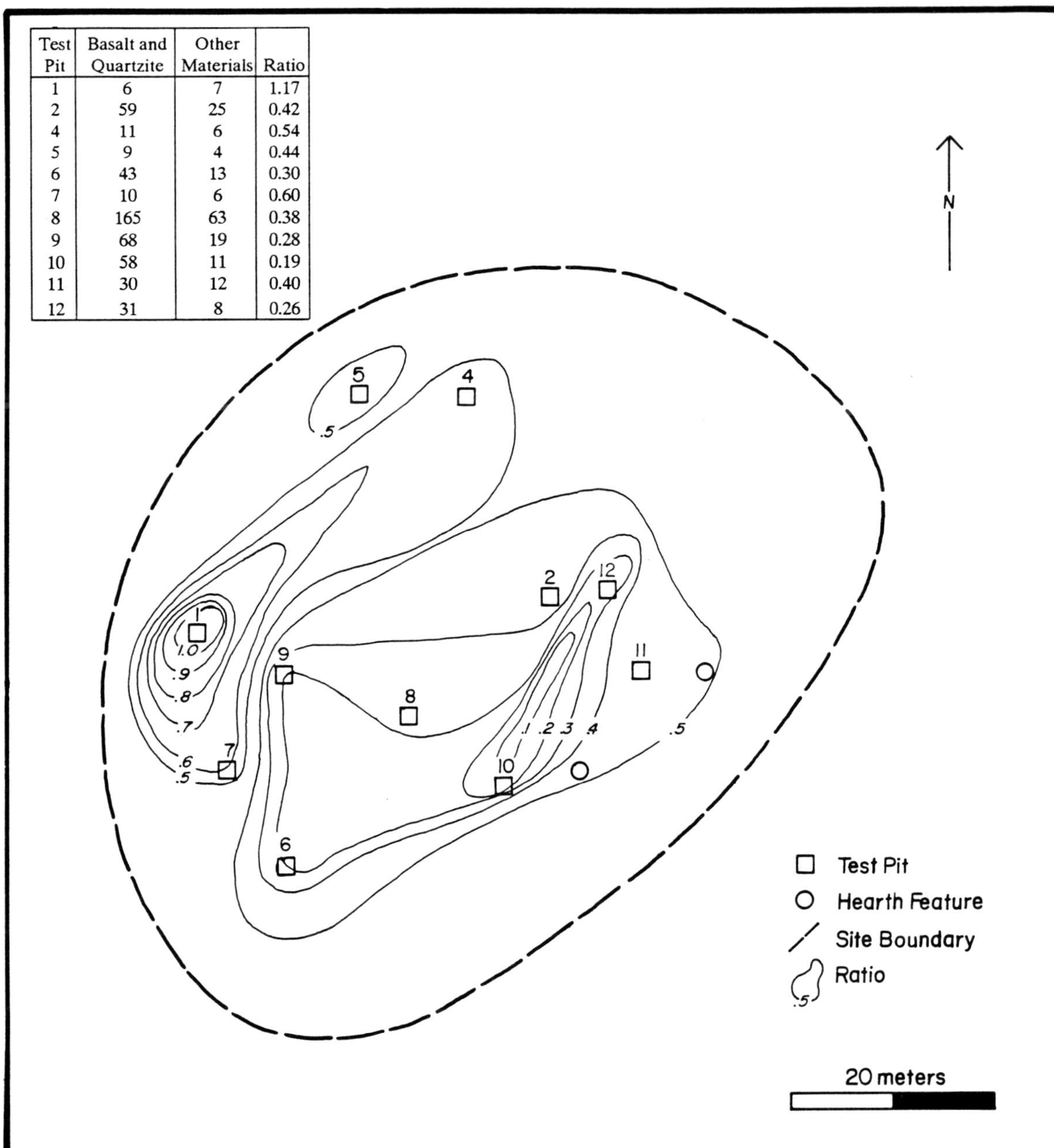

Figure 30. Ratio of basalt and quartzite to other materials.

The reduction of materials and application of technologies other than bifacial can be seen as secondary tasks within the boundaries of SDI-6010 (see Tables 17 and 21). Though evidence of these activities occurs throughout the site, one would expect it to appear in smaller ratios compared to indications of the primary activity (biface tool manufacture) within the bounds of activity areas.

Figure 31 is a distribution map based upon the ratio of shatter to flakes recovered in each test unit. The higher ratio of shatter in units 1, 9, and 4 represents a higher occurrence of artifacts produced by reduction techniques (primary and secondary reduction flakes) in this portion of the site.

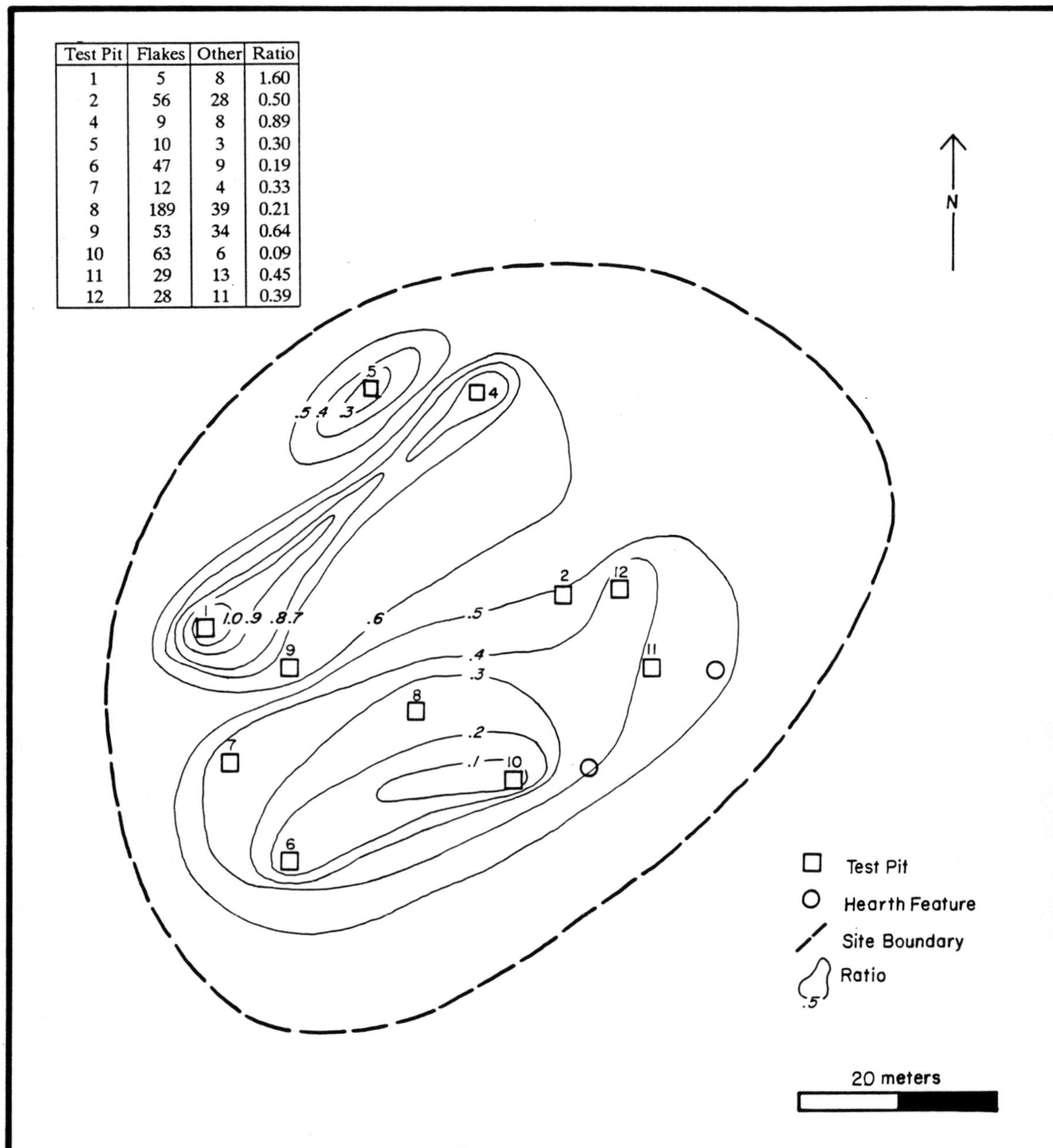

Test Pit	Flakes	Other	Ratio
1	5	8	1.60
2	56	28	0.50
4	9	8	0.89
5	10	3	0.30
6	47	9	0.19
7	12	4	0.33
8	189	39	0.21
9	53	34	0.64
10	63	6	0.09
11	29	13	0.45
12	28	11	0.39

Figure 31. Ratio of shatter to flakes.

Test Pit 5 to the north, and the rest of the units south of this area all display low ratios of shatter indicating more flakes were produced in these areas. The area of highest flake production corresponds to the lowest ratio south of the site center and just outside the cluster of hearth features.

The ratio of flakes produced by other technologies to bifacial flakes within each test unit is the basis for Figure 32. Here again we find the center of primary activity to be south of the site center, with areas to the north exhibiting a higher ratio of artifacts from secondary activities.

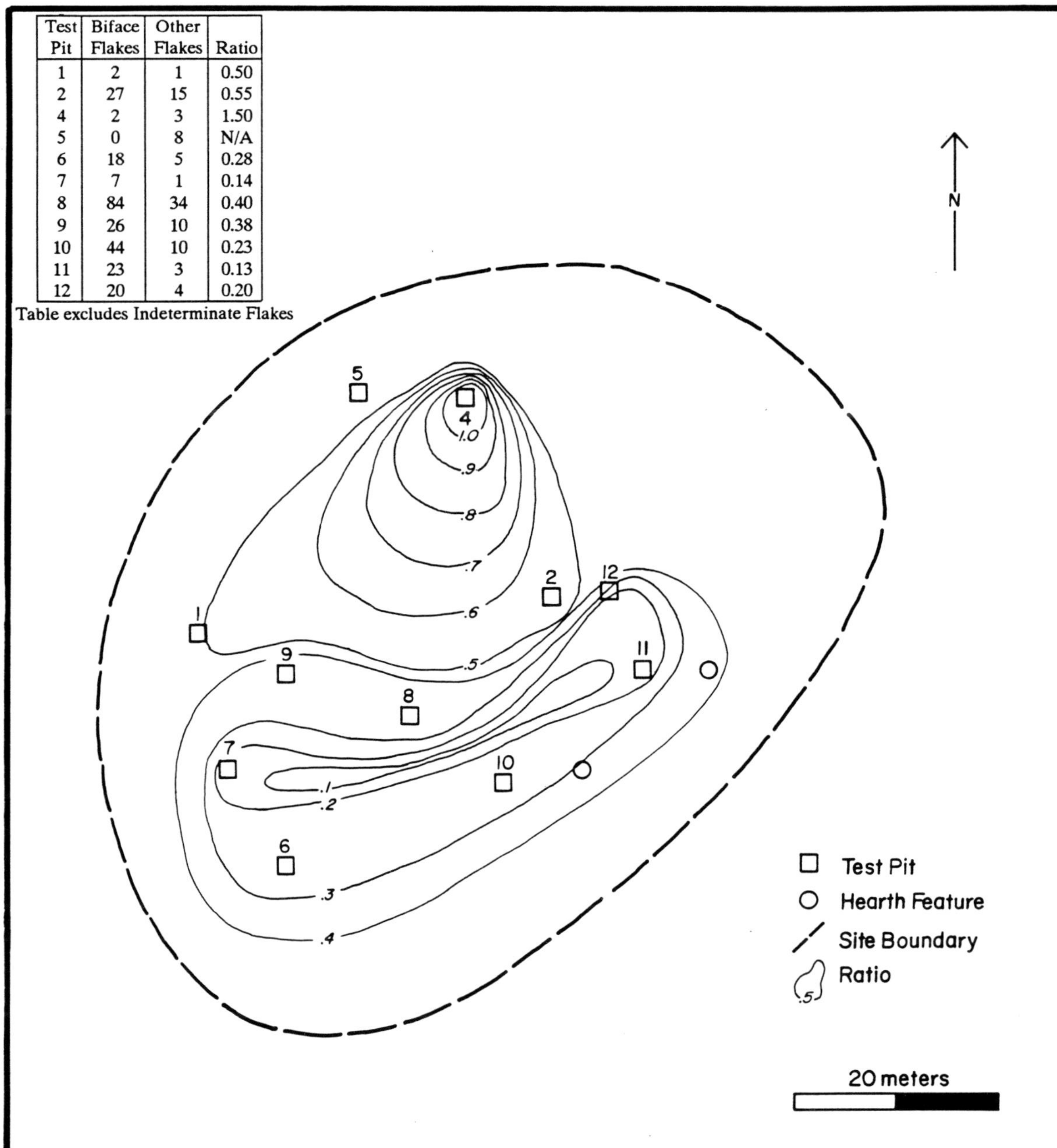

Test Pit	Biface Flakes	Other Flakes	Ratio
1	2	1	0.50
2	27	15	0.55
4	2	3	1.50
5	0	8	N/A
6	18	5	0.28
7	7	1	0.14
8	84	34	0.40
9	26	10	0.38
10	44	10	0.23
11	23	3	0.13
12	20	4	0.20

Table excludes Indeterminate Flakes

Figure 32. Ratio of other types of flakes to biface flakes.

Although there is a fair amount of variation between Figures 26 through 29, each clearly suggests the presence of a large, primary activity area within the southeastern portion of SDI-6010. Figures 27 and 28 also indicate the possible existence of a smaller activity area in close proximity to Test Pit 5. The large activity area appears to orient the site toward the San Luis Rey River approximately 200 m to the southeast of the site. Clearly, the utilization and exploitation of the river and its resources was an important aspect of adaptations at SDI-6010.

The small size of the lithic assemblage and limited amount of excavation at SDI-6010 make it difficult to reconstruct site dynamics. Further field investigation would be necessary to better define the extent of the activity area that appears to be associated with the hearth features in the southeast quadrant and to determine if any other activity areas exist.

Flaked Stone Conclusions

Although the flaked stone assemblage from the Whelan Lake site is relatively small (N=704), it contains a substantial amount of information concerning the exploitation and utilization of lithic materials at the site. Eight raw material types were utilized. The most easily accessible of these appears to have been basalt, quartzite, granite and quartz/quartz-crystal. The first three were available along the coast in the form of cobbles where limited numbers of fine-grained basalt cobbles may also have been present. Quartz/quartz-crystal was accessible in natural outcroppings occurring in the general area of the site. Cryptocrystalline silicates and Piedra del Lumbre chert were available as outcroppings in the Camp Pendleton area some 10 km to 20 km north of the site. The obsidian was no doubt obtained through trade, as its nearest source to the site is several hundred kilometers distant.

Examination of the debitage in the assemblage clearly indicates that the primary objective of lithic procurement and reduction at the site was the manufacture of biface implements. It appears that raw materials (with the exception of quartz/quartz-crystal) were, for the most part, brought to the site as previously reduced cores and blanks or as finished tools. There is evidence to suggest some expedient tools were manufactured on site from basalt and quartzite. Other tools utilized at the site may have been imported as finished or nearly finished implements.

The small number of biface implements recovered at SDI-6010 indicates that the tools manufactured at the site were transported elsewhere. The small number of cores and absence of cores of materials other than basalt and quartzite indicates the curation of raw materials probably took place elsewhere. It is also possible that both cores and biface implements have been removed from the site by various groups over the passage of time.

For the most part, the exploitation of materials at the site was similar. Due to the availability of basalt, quartzite and quartz/quartz-crystal, these materials were exploited somewhat less economically than CCS, PDL chert and the exotic obsidian. The majority of debitage in all material types reflects the application of bifacial technology. The presence of flake/core reduction and bi-polar technologies is minimal. The emphasis upon biface manufacture observed in the debitage suggests the site was utilized in part as a "gearing up" area (Binford 1979) associated with a hunting technology.

The exploitation of lithic raw materials at Whelan Lake reflects only one aspect of a complex network of technologies and materials utilized by the prehistoric inhabitants of northwestern San Diego County. The site was probably utilized only for relatively short periods of time, perhaps on a seasonal basis, and it appears only certain, specific tasks were performed there. Thus, the use of a curated lithic technology offers advantages. The tools and materials exploited at the site appear to have been both utilized and discarded elsewhere.

GROUND STONE

The ground stone assemblage from SDI-6010 comprises only 15 items: one possible metate fragment, five complete manos, and nine unidentifiable ground stone fragments. In addition, three items classed as chipped stone were identified by D.L. True (personal communication 1993) as also containing ground surfaces. All the ground stone items are formed of granite. Three of the items were collected from backhoe trench backdirt piles and are not provenienced. One mano was surface collected. Table 22 presents the data for all ground stone items recovered during testing of SDI-6010.

The possible metate fragment is a large, fractured block of granite exhibiting a concave grinding surface on one face. It is otherwise undiagnostic. It is assumed metates in conjunction with manos were used at SDI-6010 primarily to process vegetable and other food matter, but the type and extent of such use cannot be determined.

Two manos, ground on both faces and meticulously shaped, were recovered from near Feature 1 in Test Pit 2 (Figure 33). They display no signs of hammering or pecking and are assumed to have been used in processing vegetable and other food matter. The three other manos in the assemblage were recovered from trench backdirt. These items are cobbles showing grinding wear on only one surface. Like the manos associated with Feature 1, they show no signs of being used as hammers and are assumed to have been used in the processing of food matter.

Table 22. Ground Stone Distribution.

Test Unit	Level (cm)	Raw Material	Type	Status	Length	Width	Thickness	Comments
2		Granite	Unknown	Fragment				formerly classed as chipped stone
2	20-30	Granite	Unknown	Fragment				
2	20-30	Granite	Unknown	Fragment				
2	20-30	Granite	Unknown	Fragment				
2	20-30	Granite	Mano	Complete	13.53	11.84	6.27	ovoid, bifacial shaped
2	20-30	Granite	Mano	Complete	14.37	11.06	5.73	ovoid, bifacial shaped
4	10-20	Granite	Unknown	Fragment				
8	50-60	Granite	Unknown	Fragment				
8	50-60	Granite	Metate	Fragment				
9		Granite	Unknown	Fragment				also chipped stone
9		Granite	Unknown	Fragment				also chipped stone
12	30-80	Granite	Unknown	Fragment				
12	0-30	Granite	Unknown	Fragment				
12	0-30	Granite	Unknown	Fragment				
P.P.		Granite	Mano	Complete	10.82	8.82	5.41	unifacial unshaped, from surface
N.P.		Granite	Unknown	Fragment				from Trench A
N.P.		Granite	Mano	Complete	10.43	8.03	4.18	unifacial unshaped, from Trench B
N.P.		Granite	Mano	Complete	14.15	10.36	6.03	unifacial unshaped, from Trench B

KEY:N.P.=Non Provenienced; P.P.=Point Provenienced

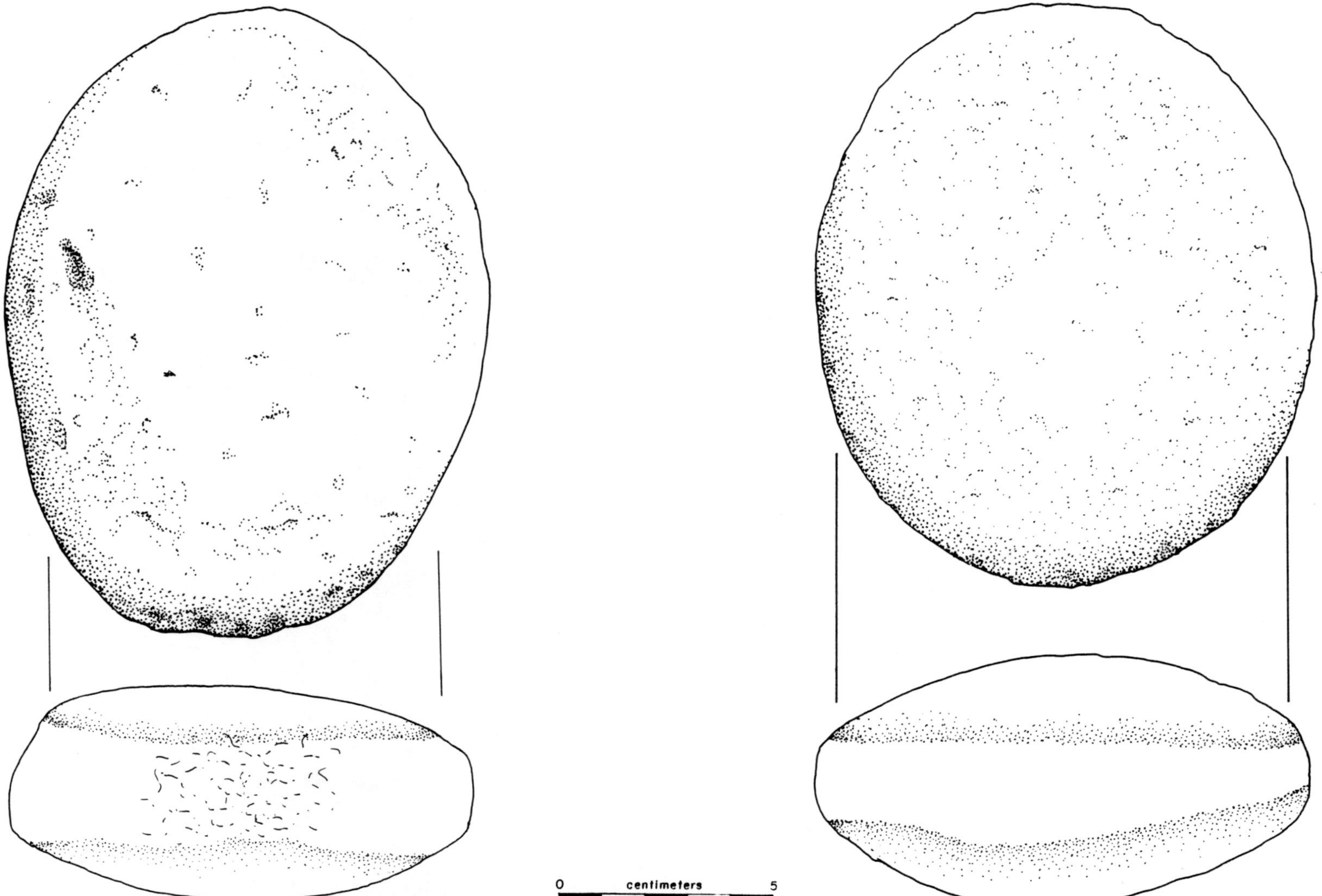

Figure 33. Shaped Manos.

Ground Stone Conclusions

Clearly, ground stone technology was utilized at SDI-6010, but the extent is difficult to ascertain. The lack of metates in the assemblage suggests that the utilization of ground stone was limited. The three cobbles exhibiting grinding wear on only one surface demonstrate the application of an expedient approach to the processing of food matter, yet the presence of a more formal, curated technology is represented by the two shaped manos associated with Feature 1 and the pic recovered from the back dirt along Trench A which was most likely used in shaping ground stone tools. In light of the small size of the assemblage, it is likely that the vegetable and other food matter utilized at SDI-6010 did not require much high-energy processing. In the absence of a larger ground stone assemblage, no positive assertions concerning the application of ground stone technology are possible.

CHAPTER 8
INVERTEBRATE FAUNAL ANALYSIS

C. Kimberly Greene McClure and Jeffrey H. Altschul

The majority of material recovered from SDI-6010 consists of invertebrate faunal remains, particularly shellfish remains. This chapter presents the results of an analysis designed to describe the collection and place the results in a regional context. Four major foci structured this analysis. These are:

(1) Identification: What shellfish species are present at SDI-6010? In what proportions are these species represented?

(2) Habitat: From what riparian and estuarine ecological zones do the various shellfish species from SDI-6010 originate?

(3) Intrasite Comparison: Is there variability in the proportional representation of shellfish species across the site? Is there evidence of shellfish processing in the features?

(4) Intersite Comparison: How does the assemblage from SDI-6010 compare with those of other sites in the region? In what role in the settlement system did SDI-6010 function?

METHODS

Shell was recovered from 10 1 m x 1 m and two 2 m x 2 m test pits. One of the 2 m x 2 m test pits, Test Pit 2, contained a hearth feature, as did one of the 1 m x 1 m test pits, Test Pit 12. Because we wanted to determine whether shellfish processing was an activity associated with these features, the shell recovered from these test pits was analyzed completely. To provide a control for this analysis as well as increase our spatial coverage, we also analyzed the shell recovered from two 1 m x 1 m test pits, Test Pits 6 and 9, located in other portions of the site. Although not representative in a statistical sense, the analyzed collection comprises 40 percent of the total number of shell hinges as well as 33 percent of the total surface area excavated (Table 23).

In the field, shell was recovered by dry-screening test pit and feature fill through 1/8 inch (3.175 mm) hardware cloth screens. As with all other material categories, shell was provenienced by test pit and 10 cm arbitrary level. Shell was initially sorted in the field to separate shell fragments larger than approximately 12 mm (thumbnail size). Shell fragments smaller than 12 mm were not collected due to the difficulty in identifying them.

Upon delivery to the laboratory, the shell was cleaned in warm water by lightly scrubbing away dirt and encrustations with a soft brush. Identification of the various shellfish species was guided by a number of handbooks and references (for example, Abbott 1968; Cerreto and Foertsch 1985; McLean 1978; Rehder 1981) as well as a type collection housed at Statistical Research.

Table 23. Total Shell NISP by Test Pit.

	TEST PIT												
	1	2	3	4	5	6	7	8	9	10	11	12	Total
Total Shell Hinges	26	537	0	126	102	363	452	827	131	359	106	327	3356

Two measures were calculated to aid in the analysis of the collection. The first is the minimum number of individuals (MNI). As with any type of faunal analysis, MNI of shell can be viewed as a standardized measure of the importance of various shell species. By focusing only on one element, the hinge, MNI measures tend to lower the seeming importance of large mollusc species, for which one shell may be broken into numerous pieces, and increase the proportional representation of smaller, more compact species.

At SDI-6010, MNI was calculated by counting all shell containing more than half of the umbo (or hinge). Because the majority of shellfish species were bivalves, this number was divided by two to obtain the MNI. The MNI of gastropods was determined by counting the number of fragments that represented more than half of the shell.

MNI measures are highly sensitive to the size of the provenience unit from which it is based. For example, if in level 1 of a 2 m x 2 m unit we find four umbos from the same species, we would calculate an MNI of 2 (4 divided by 2). In contrast, if the 2 m x 2 m unit was divided into four 1 m x 1 m units, each containing one umbo, then for the same level we would have an MNI of 4. At SDI-6010, our primary purpose for measuring MNI was to compare the horizontal and vertical distribution of shellfish species. We wanted to know if certain species were restricted to specific levels in the midden or particular parts of the site. For this purpose, we decided to calculate MNI for each level in each test pit. Because we used both 2 m x 2 m and 1 m x 1 m test pits, a slight bias enters the comparisons, with the 2 m x 2 m test pits being slightly underestimated after we control for size. Even so, the MNI represent reasonably comparable measures.

A second measure taken for each provenience was the total weight of shell. This measure provides the counterbalance to the MNI, for it allows heavy, large pieces and numerous fragments from the same shell to count equally. Weight was measured using an Ohaus Dial-O-Gram balance with an accuracy of ± 0.1 g.

SHELLFISH IDENTIFICATION AND HABITAT RECONSTRUCTION

The four test pits analyzed yielded a total of 8,309.62 g. Twenty-seven species of shellfish were identified (Table 24). The test pits averaged 308.42 individuals per cubic meter, with Test Pit 6 containing the highest concentration, 498 individuals per m^3, and Test Pit 9 the least, 170 individuals per m^3.

All shell recovered from SDI-6010 was marine in origin. This fact is of interest since the site lies approximately 7 km from the ocean. One might suggest that the presence of marine shell at SDI-6010 indicates that an estuarine lagoon was located nearby during the time the site was occupied. The close proximity of the San Luis Rey River is suggestive of the composition of an estuarine delta, a coastal inlet that has filled with sediment. Broad belts of intertidal mud flats are likely to occur in all estuarine deltas, and the areal range of open water is restricted.

Table 24. Shellfish Species Identified at SDI-6010.

Scientific Name	Common Name	Habitat
Americardia biangulata		Sublittoral gravel near rock
Argopecten sp.	Scallop	Sandy or muddy bottoms of bays and lagoons below low tide level
Barnacle	Barnacle	
Chione californiensis	Common Chione	Sandy mudflat areas in bays and sloughs, close to the surface, also offshore
Chione fluctifraga	Smooth Chione	Mud and sand flats in back bays
Chione undatella	Wavy Chione	Bays and lagoons, chiefly in sandy areas, also offshore
Donax gouldii	Bean Clam	Muddy areas of bays and on offshore mud bottoms
Crepidula onyx	Onyx Slipper Shell	Bays and lagoons
Epitonium tinctum	Tinted Wentletrap	Near the base of small sea anemones exposed at low tide
Haliotis cracherodii	Onyx Slipper Shell	Crevices or on the undersides of large rocks from the mid-tide level to low water
Irus lamellifer		Rock cavities and pholad holes, at low tide and in the sublittoral zone
Laevicardium substriatum	Egg Cockle	Sand bottoms of sloping banks at low tide and offshore
Luciniscа nuttalli	Nuttall's Lucine	Sandflats at low tide and offshore
Ostrea lurida	Native Oyster	Bays, attached to rocks, cement pilings, of oyster shell on mudflats
Petricola carditoides	Petricola carditoides	Empty pholad holes in soft rock, intertidally and offshore
Polinices reclusianus	Recluz's Moon Snail	Lagoons and shallow bays
Protothaca laciniata	Rough-sided Littleneck	Sand of sandy mud in bays or in shallow water offshore, living several inches below the surface
Protothaca staminea	Common Littleneck	Sandy areas in bays and in coarse gravel under loose rocks along the open coast, middle to lower intertidal zones
Protothaca tennerima	Thin-shelled Littleneck	Sandy mud of bays and offshore
Saxidomus nuttalli	Washington Clam	Sandy areas of bays at low tide and in sand near rocky areas along the open coast
Stenoplax conspicua	Conspicuous Chiton	Rocks resting in sand
Tagelus californianus	California Jackknife Clam	Sandy mudflats of bays
Tivela stultorum	Pismo Clam	Broad sandy beaches exposed to strong surf
Ventricolaria fordii		Gravel beds adjacent to rocky areas under kelp

The possibility of an ancient lagoon extending from the ocean to SDI-6010 was examined by Orme (see Chapter 2). He concluded that no evidence exists to support this contention. Instead, the geomorphic data available for the the San Luis Rey River basin strongly suggest that until about 5,000 B.P. a restricted estuarine lagoon existed near the present mouth of the river, extending at most 2 km inland from the coast. This lagoon probably was present when the site was occupied. Even so, at the maximum extent of the lagoon, the site was over 5 km from it.

How, then, did the residents of SDI-6010 obtain shellfish? The most logical explanation is that they traveled down river, either by boat or on foot, collected shellfish and returned to the site. If SDI-6010 was situated on the edge of a lagoon, then one might hypothesize that the residents would focus their energies on collecting shellfish close by. The proportional representation of shellfish species in the collection, then, might provide a clue as to the environmental setting of the site. By virtue of not living right by the resource, however, we can presume that the residents exercised some degree of discretion over the parts of the lagoon from which to collect shellfish. The proportional distribution of shellfish in the collection, then, can be taken as a proxy of the habitats exploited by the prehistoric inhabitants of SDI-6010.

Coastal environments can be divided into six habitats: rocky intertidal, rubble reefs, sandy beaches, bays and lagoons, rocky subtidal, and sand and mud bottoms offshore (McLean 1978). These habitats occur due to amount of light, tide, and water depth. Most species prefer one particular habitat, but can live in others, depending on environmental fluctuations. Further, it is not uncommon for more than one species to inhabit any given ecological zone. A relatively common occurrence, which is still only poorly understood, is for one species of shellfish to be predominant in a particular ecological zone one year, and replaced the next by a completely different species. Even with these radical shifts it is possible to study prehistoric resource procurement behavior, if not food preference.

The following are brief descriptions of the six ecological zones, as specified by McLean (1978).

(1) Rocky Intertidal: The most productive area in numbers of shellfish species. Littorinas and limpets occur in the splash zone (uppermost intertidal). Exposed mussel beds are common. Many small species can be found in the lower tide zone, associated with eelgrass (*Phyllospadix*). In areas where soft shale is exposed, boring clams can be found.

(2) Rubble Reefs: Low tides expose rocky beaches with large cobbles and small boulders, where fine sand accumulates between the rocks, producing a habitat suitable for chitons, but little else. This zone was probably not used by the inhabitants of SDI-6010, because the shifting sand levels are not a dependable environment for a marine shell habitat.

(3) Sandy Beaches: Although sandy beaches are generally not a productive habitat, they do support a few species, including Pismo Clam (*Tivela stultorum*) and Bean Clam (*Donax gouldii*), both of which are present at SDI-6010. Sometimes other species are washed ashore after storms.

(4) Bays and Lagoons: This ecological zone produces several species. *Cerithidea* and two pulmonate snails (none of which were present at SDI-6010) are common at the drift line in back bays. Other snails occur at low tide on the sand and mud flats. Clams burrow beneath the surface. Mussels and oysters can be attached to rocks and pilings, along with some predatory snails. Rocky channels and entrances to bays produce planktonic foods for sessile species (animals incapable of moving voluntarily, and usually attached to another organism), which support most of the predatory species.

(5) Rocky Subtidal: This area is an extension of the lower intertidal zone and can be collected only by diving. Species that are rarely exposed at low tide are generally common in the sublittoral zone. The people occupying SDI-6010 would have had access to this zone only by diving or, on a rare occasion, collecting specimens that had washed ashore.

(6) Sand and Mud Bottoms Offshore: Species populating this zone are at depths well beyond the influence of the surf and are generally collected by dredging. It is highly unlikely that the residents of SDI-6010 frequented this zone for the procurement of shellfish.

Table 24 indicates that the residents of SDI-6010 focused their shellfish procurement activities on the mudflats of the lagoon at the mouth of the San Luis Rey River, with lesser, but still considerable, attention spent on the sandy beaches of the coast. These observations come into sharper focus when we note that of the 27 species identified at SDI-6010, five account for 85.61 percent of the total shell weight (includes unidentifiable shell), and 94.35 percent of total identifiable shell weight. These five valves appear to be the economic species sought by the residents of SDI-6010, with most of the other species represented in the assemblage, collected as adjuncts to the procurement of the preferred species. Because of their presumed economic importance, the habitat and range of the dominant species from SDI-6010 are described below.

Pacific Calico Scallop (*Argopecten circularis*)

Habitat: On sand and mud, in water 10-150 feet (3-46 meters) deep.
Range: Santa Barbara, California to Peru

Common Chione (*Chione californiensis*)

Habitat: Found in the sand and mud flats of shallow bays and lagoons near the low tide line. Although less common in deep water, Common Chione also occurs in waters up to 200 feet deep. It also commonly grows in large numbers near the surface of muddy and sandy deposits.
Range: Santa Barbara, California to Panama

Smooth Chione (*Chione fluctifraga*)

Habitat: A fairly common species found in intertidal mud flats and sand of bays and lagoons.

Range: Northern California to Baja California

Wavy Chione (*Chione undatella*)

Habitat: Found in the intertidal mud flats of bays and lagoons. Wavy Chione commonly occurs in shallow waters, areas where they are easy to procure.
Range: Santa Barbara, California to Peru

Bean Clam (*Donax gouldii*)

Habitat: Abundant in some years along exposed sandy beaches along the open coast.
Range: Santa Cruz County, California to southern Baja California, sporadic north of Point Conception

The evidence indicates a clear preference selecting shellfish from the mudflats of the lagoon. It is important to point out several logical explanations for this fact. First, it is possible that the inhabitants of SDI-6010 preferred clams and scallops to other shellfish. Second, at the time the site was occupied, the mudflat environment may have predominated in the lagoon, perhaps due to a short-term environmental event, such as a flood, or to long-term environmental trends, such as continued sedimentation. Other environments, such as the rocky intertidal, may have been submerged, with their associated species correspondingly suppressed. Third, other environments of the coast may have been equally exploited, but only valves from the mudflats were brought back to the site. Economic shellfish species, such as oyster and littleneck, may have been consumed at sites closer to the lagoon, with only clam and scallop, which would have been found closest to the mouth of the river, transported back to SDI-6010.

At this time we cannot confirm or reject any of these hypotheses. Further work at sites dating to the same time period as SDI-6010 and located closer to the lagoon are needed to provide critical data.

INTRASITE ANALYSIS

Because the hearth features at SDI-6010 reflect a specific set of activities, whereas the judgementally placed test pits represent general midden, we hypothesized that proportional differences in shellfish between test pits with features and those without could be the result of dissimilar utilitarian roles of the units. Specifically, we wanted to test the notion that the features were involved in shellfish processing. Kroeber (1925) provides ethnographic accounts of bean clam (*Donax gouldii*) processing in which the gastropod was boiled in hearths to create a soup or stew. This practice might leave tangible evidence in the archaeological record in the form of disproportionately high frequencies of bean clam shell in features in relation to other areas of the site. The first goal of the intrasite analysis, then, was to determine if there was a statistical difference in the shell collections between the test pits and features.

Features 1 and 2 were found at the base of Test Pits 2 and 12, respectively. The stratigraphy of Test Pits 2 and 12 showed no indication that the upper levels of the test pits were related to the features. It is possible, then, that the shellfish recovered from the fill of Test Pits 2 and 12 is not substantially different in nature than those obtained from any other portion of the midden. In that case, the four test pits could be viewed as distinct samples of midden from various portions of the site. The second goal of the analysis was to analyze the vertical and horizontal distribution of shellfish remains from the four test pits to determine if significant changes in the use of various species occurred over time or space.

Prior to undertaking such an analysis, it was critical to demonstrate that similar arbitrary levels from different test pits could be equated as meaningful analytic units. For example, is Level 3 from Test Pit 2 equivalent to Level 3 in Test Pit 6? Such a demonstration is based primarily on comparing the proportional shell weight contained in each level from one test pit to another.

In order for our first goal to be reached, data on shell from individual levels within each test pit and feature were classified so that each unit could be handled as a single analytic unit. Test Pit 2 was further divided into two parts based on field observations. Levels 1 and 2, which appeared to be above the feature, were combined as one analytic unit as were Levels 3 and 4, which were believed to be feature fill. The comparison of shellfish from feature and midden contexts was addressed by first calculating total weight and MNI values for all shell assemblages from test pits and features (Tables 25 and 26). The frequencies and proportions of individual shell species were then compared between test pits and features.

Variations in total shell weight correspond to differences in the depth and intensity of the midden. Test Pits 2 and 6 were placed in areas where the midden was the thickest, whereas Test Pits 9 and 12 were found on the midden fringe. Differences in total shell weight between the center of the midden and the edges are mirrored by similar trends in the distribution of the total number of lithic and faunal remains.

While differences in weight probably reflect the position of the units in the midden, variations in the proportion of species between test pits are more difficult to interpret. All the test pits were dominated by remains of *Chione*, but fluctuations were noted in the overall importance of *Chione* and the types of *Chione* favored. Test Pit 9 was characterized by the least diverse shellfish assemblage with 88.7 percent of the total shell weight and 93.3 percent of MNI representing either common or wavy chione. Test Pits 6 and 12 were more comparable, with species of *Chione* accounting for between 75 and 80 percent of the total shell weight. The major difference between these test pits was in relation to MNI, with Test Pit 6 comprised of about 70 percent *Chione* and just over 20 percent bean clam and Test Pit 12 consisting of less than 40 percent *Chione* and over 50 percent bean clam.

The differences in the proportion of *Chione* and bean clam are consistent with the hypothesis that the hearths at SDI-6010 were used in part for the processing of bean clam. This hypothesis was given further support by the results from Test Pit 2. Non-feature fill (Levels 1 and 2) from Test Pit 2 is dominated by smooth chione, whereas feature fill (Levels 3 and 4) is more evenly divided between common and wavy chione, with smooth chione representing less than 5 percent by either weight or MNI. Further differentiating the two fill units was the fact that the non-feature fill contained no bean clam, while this species comprised over 35 percent of the MNI from inside the feature. Test Pit 2, therefore, appears to have been formed from two depositional events. The first was associated with Feature 1, in which bean clam was processed along with the consumption of common and wavy chione. At a later date, a second depositional event occurred, characterized by the apparent consumption of smooth chione.

While Tables 25 and 26 appear to support the link between bean clam and hearths, we wanted a more objective measure of this relationship. To this end, we computed a chi-square test between the MNI of bean clam versus the MNI for all other identifiable shellfish species for feature and non-feature contexts. The chi-square test yielded a score of 9.78, which is significant at an alpha level greater than 0.01 (Table 27). This score indicates that the likelihood of obtaining the observed shellfish distribution due to chance alone is less than 1 in a 100. Thus, fill from Features 1 and 2 contains significantly higher proportions of bean clam than general midden contexts. This association is consistent with ethnographic accounts linking bean clam processing with hearths. Establishing the link is a first step. Future research needs to focus on the exact steps in this activity, and whether bean clam processing represents the only activity associated with these hearths or is only one of many.

Table 25. Total Shell Weight (in g) by Test Pit and Feature.

	TEST PIT											
	2		2		6		9		12		TOTAL	
	Level 1&2		Level 3&4									
SPECIES	N	%	N	%	N	%	N	%	N	%	N	%
Americardia biangulata	0.00	0.00	8.40	0.51	0.00	0.00	0.00	0.00	0.00	0.00	8.91	0.11
Argopecten circularis	18.30	1.50	155.00	9.36	190.00	6.25	60.30	4.90	89.49	7.86	513.09	6.17
Barnacle	0.00	0.00	0.00	0.00	0.00	0.00	0.00	0.00	5.60	0.49	5.60	0.07
Chione californiensis	251.20	20.58	497.50	30.03	596.90	19.63	204.00	16.57	254.54	22.37	1804.14	21.71
Chione fluctifraga	618.30	50.65	88.50	5.34	43.20	1.42	0.00	0.00	28.71	2.52	778.71	9.37
Chione undatella	176.40	14.45	621.50	37.52	1670.00	54.93	889.00	72.19	533.10	46.85	3890.00	46.81
Crepidula onyx	0.00	0.00	5.50	0.33	39.40	1.30	0.00	0.00	0.00	0.00	44.90	0.54
Donax gouldii	0.00	0.00	41.00	2.48	0.00	0.00	0.01	0.00	49.10	4.31	90.11	1.08
Epitonium tinctum	0.00	0.00	5.80	0.35	0.00	0.00	0.00	0.00	0.00	0.00	5.80	0.07
Haliotis cracherodii	2.10	0.17	0.00	0.00	0.00	0.00	0.00	0.00	0.00	0.00	2.10	0.03
Haliotis sorenseni	0.00	0.00	0.00	0.00	4.10	0.13	0.00	0.00	0.00	0.00	4.10	0.05
Irus lamellifer	0.00	0.00	6.30	0.38	0.30	0.01	0.00	0.00	8.70	0.76	15.30	0.18
Laevicardium substriatum	0.00	0.00	0.00	0.00	0.00	0.00	0.00	0.00	6.30	0.55	6.30	0.08
Lucinisca nuttalli	0.00	0.00	7.10	0.43	0.00	0.00	0.00	0.00	0.00	0.00	7.10	0.09
Mactra californica	0.00	0.00	0.00	0.00	0.50	0.02	0.00	0.00	0.00	0.00	0.50	0.01
Ostrea lurida	0.50	0.04	15.80	0.95	8.31	0.27	0.60	0.05	14.70	1.29	39.91	0.48
Petricola carditoides	0.00	0.00	6.60	0.40	5.20	0.17	1.00	0.08	0.00	0.00	12.80	0.15
Polinices reclusianus	39.60	3.24	0.00	0.00	0.00	0.00	0.00	0.00	0.00	0.00	39.60	0.48
Protothaca laciniata	14.60	1.20	8.40	0.51	14.20	0.47	10.70	0.87	0.00	0.00	47.90	0.58
Protothaca staminea	21.33	1.75	22.70	1.37	14.00	0.46	2.10	0.17	21.43	1.88	81.56	0.98
Protothaca tennerima	5.10	0.42	6.00	0.36	3.70	0.12	0.50	0.04	0.00	0.00	15.30	0.18
Saxidomus nuttalli	4.20	0.34	19.30	1.17	1.40	0.05	6.60	0.54	0.00	0.00	31.50	0.38
Stenoplax conspicua	0.00	0.00	0.90	0.05	0.00	0.00	0.00	0.00	0.00	0.00	0.90	0.01
Tagelus californianus	0.00	0.00	5.90	0.36	1.90	0.06	0.00	0.00	12.10	1.06	19.90	0.24
Tivela stultorum	0.00	0.00	14.10	0.85	11.70	0.38	1.50	0.12	22.90	2.01	50.20	0.60
Tresus nuttalli	0.00	0.00	0.00	0.00	0.60	0.02	0.00	0.00	0.00	0.00	0.60	0.01
Ventricolaria fordii	0.00	0.00	0.00	0.00	0.10	0.00	0.00	0.00	5.70	0.50	5.80	0.07
UNIDENTIFIED	69.00	5.65	120.20	7.26	434.70	14.30	55.10	4.47	108.50	9.53	787.50	9.48
TOTAL	1220.63	100.00	1656.50	100.00	3040.21	99.99	1231.41	100.00	1160.87	102.01	8309.62	100.00

Table 26. MNI by Test Pit and Feature.

	Test Pit											
	2		2		6		9		12		Total	
	Level 1&2		Level 3&4									
SPECIES	N	%	N	%	N	%	N	%	N	%	N	%
Americardia biangulata	0.00	0.00	*	0.00	0.00	0.00	0.00	0.00	0.00	0.00	0.00	0.00
Argopecten circularis	2.00	2.78	10.00	5.95	15.00	4.63	6.00	5.04	6.00	3.77	39.00	4.63
Barnacle	0.00	0.00	0.00	0.00	0.00	0.00	0.00	0.00	1.00	0.63	1.00	0.12
Chione californiensis	18.00	25.00	50.00	29.76	125.00	38.58	29.00	24.37	24.00	15.09	246.00	29.22
Chione fluctifraga	40.00	55.56	8.00	4.76	1.00	0.31	0.00	0.00	2.00	1.26	51.00	6.06
Chione undatella	8.00	11.11	35.00	20.83	104.00	32.10	82.00	68.91	36.00	22.64	265.00	31.47
Crepidula onyx	0.00	0.00	*	0.00	0.00	0.00	0.00	0.00	0.00	0.00	0.00	0.00
Donax gouldii	0.00	0.00	59.00	35.12	70.00	21.60	*	0.00	85.00	53.46	214.00	25.42
Epitonium tinctum	0.00	0.00	1.00	0.60	0.00	0.00	0.00	0.00	0.00	0.00	1.00	0.12
Haliotis cracherodii	1.00	1.39	0.00	0.00	0.00	0.00	0.00	0.00	0.00	0.00	1.00	0.12
Haliotis sorenseni	0.00	0.00	0.00	0.00	1.00	0.31	0.00	0.00	0.00	0.00	1.00	0.12
Irus lamellifer	0.00	0.00	*	0.00	*	0.00	0.00	0.00	*	0.00	0.00	0.00
Laevicardium substriatum	0.00	0.00	0.00	0.00	0.00	0.00	0.00	0.00	*	0.00	0.00	0.00
Lucinisca nuttalli	0.00	0.00	1.00	0.60	0.00	0.00	0.00	0.00	0.00	0.00	1.00	0.12
Mactra californica	0.00	0.00	0.00	0.00	*	0.00	0.00	0.00	0.00	0.00	0.00	0.00
Ostrea lurida	1.00	1.39	3.00	1.79	3.00	0.93	*	0.00	2.00	1.26	9.00	1.07
Petricola carditoides	0.00	0.00	*	0.00	1.00	0.31	*	0.00	0.00	0.00	1.00	0.12
Polinices reclusianus	1.00	1.39	0.00	0.00	0.00	0.00	0.00	0.00	0.00	0.00	1.00	0.12
Protothaca laciniata	*	0.00	*	0.00	*	0.00	2.00	1.68	0.00	0.00	2.00	0.24
Protothaca staminea	1.00	1.39	1.00	0.60	1.00	0.31	*	0.00	2.00	1.26	5.00	0.59
Protothaca tennerima	*	0.00	*	0.00	1.00	0.31	*	0.00	0.00	0.00	1.00	0.12
Saxidomus nuttalli	*	0.00	*	0.00	*	0.00	*	0.00	0.00	0.00	0.00	0.00
Stenoplax conspicua	0.00	0.00	*	0.00	0.00	0.00	0.00	0.00	0.00	0.00	0.00	0.00
Tagelus californianus	0.00	0.00	*	0.00	*	0.00	0.00	0.00	1.00	0.63	1.00	0.12
Tivela stultorum	0.00	0.00	*	0.00	1.00	0.31	*	0.00	*	0.00	1.00	0.12
Tresus nuttalli	0.00	0.00	0.00	0.00	1.00	0.31	0.00	0.00	0.00	0.00	1.00	0.12
Ventricolaria fordii	0.00	0.00	0.00	0.00	*	0.00	0.00	0.00	*	0.00	0.00	0.00
TOTAL	72.00	100.00	168.00	100.00	324.00	100.00	119.00	100.00	159.00	100.00	842.00	100.00

*Fragments Only Present

Table 27. Chi Square Matrix and Score for Bean Clam Versus Other Identifiable Shellfish Species MNI for Feature and Non-feature Fill Contexts.

	Bean Clam	Other	Total
Feature	144	183	327
Non-Feature	70	445	515
Total	214	628	842

Chi-Square = 9.7795
degrees of freedom = 1
alpha level > .01

In addition to supporting the link between bean clam and hearths, Test Pit 2 provides perhaps the best evidence for the integrity of the midden. The abrupt distinctions between the two fill units indicates that little natural disturbance in the form of plant roots or animal burrowing has affected the midden. Further, no evidence was found that plowing had substantially mixed shellfish from different strata.

These observations indicated that it might be fruitful to continue the shell analysis toward our second goal: examining vertical distributions of shellfish remains. These data are presented in Tables 28 through 31. In general, vertical distribution of shellfish appears trimodal, with peaks in Levels 1, 3, and 5. The peaks in Levels 1 and 3 are approximately of the same intensity, with the peak in Level 5 about half as strong as the others. A curious peak in smooth chione is noted in Level 2, which basically represents the non-feature fill unit in Test Pit 2. The most likely explanation for this anomaly is that it represents a single processing event or series of closely related events. With the exception of smooth chione, the other four major economic species follow the general trimodal distribution.

The vertical distribution of shellfish suggests that the entire midden followed similar patterns of formation. Intensity of use fluctuated through time, with heavy usage reflected by larger frequencies of shell remains. In general, all areas of the midden appear to be affected by contemporary events at the same time, so that Level 3 in Test Pit 6 is roughly contemporaneous with Level 3 in the other units.

The condition of shell throughout the midden is roughly the same. Little decay was noted in any of the levels, suggesting that the midden formed in a relatively short period of time.

INTERSITE ANALYSIS

Although the San Luis Rey River basin contains numerous sites, few have been excavated, and even fewer have been systematically analyzed. Given its location near the upper end of the lower basin, SDI-6010 is most comparable with sites further downstream. To date, only one such site, SDI-5130, has been excavated and analyzed. Described as a large village site, SDI-5130 is located on a terrace overlooking the north bank of the San Luis Rey River, approximately 1.5 km downstream of SDI-6010 (Carrico 1978; Quillen et al. 1984; Gibson et al. 1990). Based on radiocarbon dates, the site appears to have two major components, one dating between about 6,300 and 4,000 B.P. and another dating around 140 B.P. Most of the occupation at the site appears to have occurred during the earlier component.

Table 28. Total Shell Weight (in g) by Level.

		LEVEL								
		1	2	3	4	5	6	7	TOTAL	%
SPECIES	*Americardia biangul*	0.00	0.00	8.40	0.00	0.00	0.00	0.00	8.40	0.10%
	Argopecten circulari	123.70	45.19	98.70	78.20	91.20	52.30	23.80	513.09	6.17%
	Barnacle	5.60	0.00	0.00	0.00	0.00	0.00	0.00	5.60	0.07%
	Chione californiensi	476.80	349.84	519.80	113.40	189.10	107.40	47.80	1804.14	21.71%
	Chione fluctifraga	162.91	510.80	66.70	23.10	12.00	3.20	0.00	778.71	9.37%
	Chione undatella	949.50	616.20	957.40	328.70	608.30	291.50	138.40	3890.00	46.81%
	Crepidula onyx	0.00	0.00	5.50	0.00	0.00	0.00	0.00	5.50	0.07%
	Donax gouldii	41.50	17.00	41.20	7.31	15.60	5.60	1.30	129.51	1.56%
	Epitonium tinctum	0.00	0.00	5.80	0.00	0.00	0.00	0.00	5.80	0.07%
	Haliotis cracherodii	2.10	0.00	0.00	0.00	0.00	0.00	0.00	2.10	0.03%
	Haliotis sorenseni	0.00	0.00	4.10	0.00	0.00	0.00	0.00	4.10	0.05%
	Irus lamellifer	0.40	8.70	6.30	0.00	0.10	0.00	0.00	15.50	0.19%
	Laevicardium substri	6.30	0.00	0.00	0.00	0.00	0.00	0.00	6.30	0.08%
	Lucinisca nuttalli	0.00	0.00	7.10	0.00	0.00	0.00	0.00	7.10	0.09%
	Mactra californica	0.20	0.00	0.10	0.00	0.00	0.00	0.00	0.30	0.00%
	Ostrea lurida	12.10	6.60	10.40	6.40	2.80	1.50	0.11	39.91	0.48%
	Petricola carditoides	4.20	0.00	7.20	0.00	1.00	0.40	0.00	12.80	0.15%
	Polinices reclusianus	39.60	0.00	0.00	0.00	0.00	0.00	0.00	39.60	0.48%
	Protothaca laciniata	0.90	14.60	8.40	0.00	11.20	7.90	4.90	47.90	0.58%
	Protothaca staminea	25.96	22.70	19.80	7.00	5.40	0.70	0.00	81.56	0.98%
	Protothaca tennerim	1.60	3.50	6.30	0.00	3.30	0.60	0.00	15.30	0.18%
	Saxidomus nuttalli	0.20	4.20	20.50	1.20	2.50	2.90	0.00	31.50	0.38%
	Stenoplax conspicua	0.00	0.00	0.90	0.00	0.00	0.00	0.00	0.90	0.01%
	Tagelus californianu	7.00	5.50	5.90	0.00	0.50	0.40	0.60	19.90	0.24%
	Tivela stultorum	15.00	9.40	22.00	0.00	2.30	1.50	0.00	50.20	0.60%
	Tresus nuttalli	0.60	0.00	0.00	0.00	0.00	0.00	0.00	0.60	0.01%
	Ventricolaria fordii	5.70	0.00	0.00	0.00	0.10	0.00	0.00	5.80	0.07%
	UNIDENTIFIED	282.10	187.90	183.00	21.10	80.80	30.70	1.90	787.50	9.48%
	TOTAL	2163.97	1802.13	2005.50	586.41	1026.20	506.60	218.81	8309.62	100.00%

Table 29. Percentage of Shell Weight by Level.

		LEVEL							
		1	2	3	4	5	6	7	TOTAL
SPECIES	*Americardia biangulata*	0.00	0.00	100.00	0.00	0.00	0.00	0.00	100.00
	Argopecten circularis	24.11	8.81	19.24	15.24	17.77	10.19	4.64	100.00
	Barnacle	100.00	0.00	0.00	0.00	0.00	0.00	0.00	100.00
	Chione californiensis	26.43	19.39	28.81	6.29	10.48	5.95	2.65	100.00
	Chione fluctifraga	20.92	65.59	8.57	2.97	1.54	0.41	0.00	100.00
	Chione undatella	24.41	15.84	24.62	8.45	15.64	7.49	3.55	100.00
	Crepidula onyx	0.00	0.00	100.00	0.00	0.00	0.00	0.00	100.00
	Donax gouldii	32.04	13.13	31.81	5.64	12.05	4.32	1.00	100.00
	Epitonium tinctum	0.00	0.00	100.00	0.00	0.00	0.00	0.00	100.00
	Haliotis cracherodii	100.00	0.00	0.00	0.00	0.00	0.00	0.00	100.00
	Haliotis sorenseni	0.00	0.00	100.00	0.00	0.00	0.00	0.00	100.00
	Irus lamellifer	2.58	56.13	40.64	0.00	0.65	0.00	0.00	100.00
	Laevicardium substriatum	100.00	0.00	0.00	0.00	0.00	0.00	0.00	100.00
	Lucinisca nuttalli	0.00	0.00	100.00	0.00	0.00	0.00	0.00	100.00
	Mactra californica	66.67	0.00	33.33	0.00	0.00	0.00	0.00	100.00
	Ostrea lurida	30.31	16.53	26.06	16.04	7.02	3.76	0.28	100.00
	Petricola carditoides	32.81	0.00	56.25	0.00	7.81	3.13	0.00	100.00
	Polinices reclusianus	100.00	0.00	0.00	0.00	0.00	0.00	0.00	100.00
	Protothaca laciniata	1.88	30.48	17.54	0.00	23.38	16.49	10.23	100.00
	Protothaca staminea	31.83	27.83	24.28	8.58	6.62	0.86	0.00	100.00
	Protothaca tennerima	10.45	22.88	41.18	0.00	21.57	3.92	0.00	100.00
	Saxidomus nuttalli	0.63	13.33	65.08	3.81	7.94	9.21	0.00	100.00
	Stenoplax conspicua	0.00	0.00	100.00	0.00	0.00	0.00	0.00	100.00
	Tagelus californianus	35.18	27.64	29.64	0.00	2.51	2.01	3.02	100.00
	Tivela stultorum	29.88	18.73	43.82	0.00	4.58	2.99	0.00	100.00
	Tresus nuttalli	100.00	0.00	0.00	0.00	0.00	0.00	0.00	100.00
	Ventricolaria fordii	98.28	0.00	0.00	0.00	1.72	0.00	0.00	100.00
	UNIDENTIFIED	35.82	23.86	23.24	2.68	10.26	3.90	0.24	100.00
	TOTAL	26.04	21.68	24.14	7.05	12.36	6.10	2.63	100.00

Table 30. MNI by Level.

		LEVEL								
		1	2	3	4	5	6	7	TOTAL	%
SPECIES	*Americardia biangulata*	0.00	0.00	*	0.00	0.00	0.00	0.00	0.00	0.00%
	Argopecten circularis	11.00	4.00	8.00	6.00	4.00	3.00	3.00	39.00	4.72%
	Barnacle	1.00	*	*	*	*	*	0.00	1.00	0.12%
	Chione californiensis	47.00	38.00	72.00	10.00	60.00	15	4.00	231.00	27.93%
	Chione fluctifraga	17.00	26.00	7.00	1.00	*	*	0.00	51.00	6.17%
	Chione undatella	61.00	36.00	57.00	31.00	47.00	23.00	10.00	265.00	32.04%
	Crepidula onyx	0.00	0.00	*	0.00	0.00	0.00	0.00	0.00	0.00%
	Donax gouldii	80.00	26.00	70.00	4.00	31.00	*	3.00	214.00	25.88%
	Epitonium tinctum	0.00	0.00	1.00	0.00	0.00	0.00	0.00	1.00	0.12%
	Haliotis cracherodii	1.00	0.00	0.00	0.00	0.00	0.00	0.00	1.00	0.12%
	Haliotis sorenseni	0.00	0.00	1.00	0.00	0.00	0.00	0.00	1.00	0.12%
	Irus lamellifer	*	*	*	0.00	*	0.00	0.00	0.00	0.00%
	Laevicardium substriatum	*	0.00	0.00	0.00	0.00	0.00	0.00	0.00	0.00%
	Lucinisca nuttalli	0.00	0.00	1.00	0.00	0.00	0.00	0.00	1.00	0.12%
	Mactra californica	0.00	0.00	0.00	0.00	0.00	0.00	0.00	0.00	0.00%
	Ostrea lurida	4.00	*	2.00	1.00	1.00	1.00	*	9.00	1.09%
	Petricola carditoides	1.00	0.00	*	0.00	*	*	0.00	1.00	0.12%
	Polinices reclusianus	1.00	0.00	0.00	0.00	0.00	0.00	0.00	1.00	0.12%
	Protothaca laciniata	*	*	*	0.00	*	1.00	1.00	2.00	0.24%
	Protothaca staminea	2.00	1.00	2.00	*	*	*	0.00	5.00	0.60%
	Protothaca tennerima	*	*	*	0.00	1.00	*	0.00	1.00	0.12%
	Saxidomus nuttalli	*	*	*	*	*	*	*	0.00	0.00%
	Stenoplax conspicua	0.00	0.00	*	0.00	0.00	0.00	0.00	0.00	0.00%
	Tagelus californianus	1.00	*	*	0.00	*	*	*	1.00	0.12%
	Tivela stultorum	*	*	1.00	0.00	*	*	0.00	1.00	0.12%
	Tresus nuttalli	1	0	0	0	0	0	0	1.00	0.12%
	Ventricolaria fordii	*	0.00	0.00	0.00	*	0.00	0.00	0.00	0.00%
	TOTAL	228.00	131.00	222.00	53.00	144.00	28.00	21.00	827.00	100.00%

* Fragments only present

Table 31. Percentage of MNI by Level.

	LEVEL							
SPECIES	1	2	3	4	5	6	7	TOTAL
*Americardia biangulata**	0.00	0.00	*	0.00	0.00	0.00	0.00	0.00
Argopecten circularis	28.21	10.26	20.51	15.38	10.26	7.69	7.69	100.00
Barnacle	100.00	0.00	0.00	0.00	0.00	0.00	0.00	100.00
Chione californiensis	19.11	15.45	29.27	4.07	24.39	6.09	1.62	100.00
Chione fluctifraga	33.33	50.98	13.73	1.96	*	*	0.00	100.00
Chione undatella	23.02	13.58	21.51	11.70	17.74	8.68	3.77	100.00
*Crepidula onyx**	0.00	0.00	*	0.00	0.00	0.00	0.00	0.00
Donax gouldii	37.38	12.15	32.71	1.87	14.49	*	1.40	100.00
Epitonium tinctum	0.00	0.00	100.00	0.00	0.00	0.00	0.00	100.00
Haliotis cracherodii	100.00	0.00	0.00	0.00	0.00	0.00	0.00	100.00
Haliotis sorenseni	0.00	0.00	100.00	0.00	0.00	0.00	0.00	100.00
*Irus lamellifer**	*	*	*	0.00	*	0.00	0.00	0.00
*Laevicardium substriatum**	*	0.00	0.00	0.00	0.00	0.00	0.00	0.00
Lucinisca nuttalli	0.00	0.00	100.00	0.00	0.00	0.00	0.00	100.00
Mactra californica	0.00	0.00	0.00	0.00	0.00	0.00	0.00	0.00
Ostrea lurida	44.45	*	22.22	11.11	11.11	11.11	*	100.00
Petricola carditoides	100.00	0.00	*	0.00	*	*	0.00	100.00
Polinices reclusianus	100.00	0.00	0.00	0.00	0.00	0.00	0.00	100.00
Protothaca laciniata	*	*	*	0.00	*	50.00	50.00	100.00
Protothaca staminea	40.00	20.00	40.00	*	*	*	0.00	100.00
Protothaca tennerima	*	*	*	0.00	100.00	*	0.00	100.00
*Saxidomus nuttalli**	*	*	*	*	*	*	*	0.00
*Stenoplax conspicua***	0.00	0.00	*	0.00	0.00	0.00	0.00	0.00
Tagelus californianus	100.00	*	*	0.00	*	*	*	100.00
Tivela stultorum	*	*	100.00	0.00	*	*	0.00	100.00
Tresus nuttalli	100	0.00	0.00	0.00	0.00	0.00	0.00	100.00
*Ventricolaria fordii**	*	0.00	0.00	0.00	*	0.00	0.00	0.00
TOTAL	27.57	15.84	26.84	6.41	17.41	3.39	2.54	100.00

* Fragments only present

Although SDI-5130 is much larger than SDI-6010, in most other aspects the two sites are very comparable. Both consist of sparse shell middens located well upstream from the mouth of the river. The shell assemblages of the two sites have pointed similarities and stark contrasts. Because it is smaller and has had far less cubic meters of excavated fill analyzed for shellfish, one would expect that SDI-6010 would be characterized by a more restricted shell assemblage than SDI-5130. Yet the reverse is true, with 27 species of shellfish found at SDI-6010, and only 20 species at SDI-5130 (Sandler 1990). Further, since SDI-6010 is farther from the lagoon than SDI-5130, we would expect that the latter would be characterized by more shellfish per cubic meter. Again, the opposite holds, with SDI-6010 containing on average 3035.5 g/m^3 as opposed to 1776.2 g/m^3 at SDI-5130 (Carrico 1978:77).

Although the shellfish assemblages at the two sites differ in density and diversity, they are strikingly similar in terms of composition. Both assemblages are dominated by species of *Chione*, with *Argopecten* following at a distant second, and no other species representing more than 1 percent of the total (Table 32). In examining Table 32, it is clear that the results from the 1990 excavations at SDI-5130 are at variance with previous analyses at that site. We believe that the difference is due to sample size. The 1978 and 1984 analyses were based on sample weights of 16,874 and 55,982.3 g, respectively, whereas the 1990 analysis was based on 3,000 g. We suspect that when the analysis of the larger collection recovered in 1991 is complete that it will correspond more closely with the 1978 and 1984 results.

At present, we are left with a set of observations that are difficult to reconcile. SDI-6010, a site that is smaller and more distant from the lagoon, is characterized by a more diverse and denser assemblage of shellfish than SDI-5130, a site interpreted as a major habitation village. Although the assemblages differ in absolute numbers of shellfish species present, they are both characterized by a focus on species of *Chione* and *Argopecten*.

We believe that the most likely explanation of these observations relates to the time of occupation. SDI-6010 was probably occupied between about 7,400 and 6,500 B.P., whereas SDI-5130 was primarily occupied later, between 6,300 and 4,000 B.P. It is likely that when the residents of SDI-6010 first reached the San Luis Rey basin, the lagoon was in existence and shellfish species had not been heavily exploited. The mudflats of the lagoon would have provided easy access to *Chione*, which appears to have been the target of these endeavors. Although particular species were targeted, the procurement strategy probably resulted in the collection of all shellfish species in the beds. Consequently, a large number of non-economic species were brought back to the site. Human occupation of the basin, and shellfish exploitation, appears to have been more-or-less continuous for the period covered by SDI-6010 and SDI-5130. Over time, then, it is possible that the mudflats became less productive, resulting in few species and smaller amounts of shellfish.

One approach to test this model would be to compare the average size of *Chione* shells at SDI-6010 with those at SDI-5130. If correct, we should expect to see a decrease in shell size.

While such an analysis must await future research, another approach is to examine similar aged sites nearby. The Windsong Shores site (SDI-10965) is located overlooking the Aqua Hedionda lagoon, just south of San Luis Rey (Gallegos 1991). The site has two components, one dated between 8,400 and 8,000 B.P. and the other around 7,000 B.P. The shellfish assemblage of the site is nearly identical to the Whelan Lake site. Twenty-five shellfish species are represented at the Windsong Shores site, with *Chione* comprising 87.00 percent of the collection, *Argopecten* representing 7.79 percent, *Donax* contributing 0.03 percent, and all other species accounting for 5.18 percent. Limited support, therefore, exists for relatively rich and diverse lagoons between 8,000 and 7,000 B.P. How the shellfish populations reacted to several thousand years of human exploitation still needs to be examined.

Table 32. Intrasite Shell Comparison, Percentage of Total for Each Site.

	CA-SDI-5130			CA-SDI-6010
	1978	1984	1990	
SPECIES				
Chione sp.	86.47	89.00	53.27	83.00
Argopecten sp.	9.90	6.00	21.35	6.00
Donax sp.	.00	.00	1.81	1.00
Other	3.63	5.00	23.57	10.00

CONCLUSIONS

The analysis of invertebrate remains at SDI-6010 has led to three basic conclusions about the prehistoric occupation. First, the occupation tended to be episodic. Three episodes of relatively intense shellfish exploitation were sandwiched between periods of lesser activity. The length and duration of the various episodes is difficult to determine. But based both on the radiocarbon dates, which indicate that the entire occupation probably lasted no more than 300 years, and the uniform quality of the shell throughout the midden, we would suggest that individual episodes lasted no more than a generation or two.

Our second conclusion relates to the processing of bean clams. Following Kroeber's (1925) ethnographic observations, we hypothesized that bean clams were boiled in a soup or stew, and therefore that a statistical association between bean clams and hearths should be found. Indeed, a statistically significant association between the two variables was found at SDI-6010. While we are confident that the association is not fortuitous, we cannot conclude that the only use of these hearths was for bean clam processing. Although bean clams are common near hearths, *Chione* remains are still the most frequent form of shellfish remains in the feature fills. A more plausible explanation is that while hearths were sometimes used for bean clam processing, other resources were processed as well. What the analysis demonstrates is not so much that hearths focused on bean clam processing, but rather that a set of specific activities associated with the features can be detected in the archaeological record.

The third conclusion relates to the placement of SDI-6010 in a regional setting and to overall subsistence strategies followed by early Holocene hunters and gatherers. Although the site is located 5 km upstream from the ancient lagoon, it is clear that residents utilized lagoonal resources on a regular basis. We suspect that the placement of the site at some distance from the resource was based on other subsistence practices. In a sense, the placement of the site can be viewed as a relative measure of the importance of lagoonal as opposed to terrestrial resources. Following Jones (1991), we would agree that SDI-6010 shows that coastal resources were not shunned by early hunters and gatherers. The site's placement, however, indicates that these resources were also not placed at the top of these people's subsistence decisions.

Were coastal resources viewed of secondary importance during this time period? Based on one site, we cannot even begin to address this question. Additional research would have to show that early sites are not located near the edge of the ancient lagoon, and research to date does not permit such an assessment. What can be concluded is that the residents of SDI-6010 were exploiting a lagoon with a high resource yield. Perhaps because the resource was so easy to exploit and readily abundant, the residents decided to place their site in an area where resources were more difficult to procure, but the return on successful ventures was great. Evaluation of this notion must await further research.

CHAPTER 9
FAUNAL ANALYSIS

Kellie M. Cairns and Jeffrey H. Altschul

INTRODUCTION

The analysis of the faunal remains collected from the Whelan Lake site (SDI-6010) was focused on two broad topics: the diet of the prehistoric inhabitants and the ecological zones they exploited for food. Encompassed within these broad topics are more specific questions of seasonality and hunting strategies. In our attempt to address these topics, it was necessary to examine the distribution of bones both vertically and horizontally over the midden as well as place the assemblage within a regional context.

The faunal analysis is organized into four sections. The first section addresses the methodology employed in the artifact processing and analytical phases. The second section provides a descriptive summary of the archaeological fauna observed at SDI-6010. The third section discusses questions of diet and subsistence practices. The chapter ends with a series of observations regarding the importance of hunting in the San Luis Rey River basin during the La Jolla period.

METHODS

Laboratory analysis commenced with carefully dry brushing all bone. After the bone had been bagged according to item and provenience numbers, an analysis form was used to record provenience, test pit number, depth, class of vertebrate, genus and species, weight, bone name and side, modification and comments. Identification was aided by use of the comparative collection located in the zooarchaeology laboratory of the Arizona State Museum.

Species identification was determined by diagnostic features of identifiable bone as well as by the size of the bone. Bone fragments were assigned to very small, small, medium, large, and very large categories of birds and mammals. These classifications were judged on the basis of relative body size with rodents (very small), lagomorphs (small), dog (medium), deer (large), and cow (very large). The birds were classified as passerine (small), hawks (medium), and ducks (large). Bone modifications were noted indicating burning, gnawing, cutting, and polishing.

After completing the identification, two measures were calculated. The first, NISP, refers to the number of bones identified to a particular species. This measure should not be used as the primary indicator of species abundance because it is sensitive to bone fragmentation. Instead, NISP should be complemented by a second measure, MNI, or the minimum number of individuals. MNI is a minimum estimate of taxonomic abundance at a site. This measure separates the most abundant elements from species into left and right components and utilizes the greater number of sided elements as a unit of calculation. This approach is useful in avoiding errors due to fragmentation but is sensitive to the effects of aggregation of the excavated units (Grayson 1984). MNI was computed at SDI-6010 by using the test pit as the analytic unit.

Due to the lack of a larger, more complete comparative collection, approximately 45 fragments were sent to Dr. Roy Salls of the Lompoc Museum for further analysis. Of the 45 specimens, 13 were identified as piscine fauna and 32 were mammalian.

THE ARCHAEOLOGICAL FAUNA

This section summarizes the characteristics of the faunal assemblage recovered from SDI-6010. The fauna are discussed according to natural environment, characteristics, and occurrence on the site. Fourteen genera of mammals, seven genera of birds, three genera of fish, three genera of Elasmobranchs and one genus of reptile were represented. Table 33 represents the list of species recovered from the site. The references to the fauna in this chapter are from Behler (1979), Burt and Grossenheider (1976), Bleitz-Sanburg (1987), Gibson et al. (1990), Gotshall (1989), Heizer and Elsasser (1980), Kroeber (1925), Salls (1988), and Udvardy (1984).

Mammals

Mammals represent 98.93 percent of the total bone mass, 98.70 percent of the total number of bones and 76.27 percent of the total MNI (Table 34). The majority of the exploited fauna are terrestrial and have a wide range of ecosystems including one represented sea mammal.

FAMILY GEOMYIDAE
Thomomys bottae

Pocket gophers are variable in size, weighing from two and one half to nine ounces. They are burrowers and are seldom seen above ground. These mammals inhabit not only loamy valleys adjacent to water or damp soils, but also occur in sandy or rocky environments. They prefer root crops and surface vegetation and are considered harmful whenever they occur in cultivated areas. Due to the relatively high abundance of gophers in the site, along with their natural instinct to burrow and aerate the soil, it is probable that the unit contents were disrupted by these creatures, and that many of their remains were deposited after the site was abandoned by humans.

LEPORIDAE
Lepus sp.

Weighing from three to seven pounds, this common "jackrabbit" occurs in populations throughout grassland deserts and open valleys among scattered vegetation. *Lepus* primarily forages on grass and consumes a considerable amount of vegetation that could be used by livestock. This tasty creature was a main source of meat for prehistoric populations. It should also be noted that the meat of the jackrabbit carries tularemia, a rabbit disease, which may have infected these populations.

It has been well-documented that native peoples usually obtained jackrabbit through rabbit drives. Communal rabbit drives had the potential of producing more pounds of animal protein with less expenditure of energy than other economic activities practiced by the Indians (Salls 1988).

Sylvilagus sp.

Members of the cottontail family are approximately 12 to 15 inches long. They inhabit open plains, foothills and low valleys with grass, sagebrush, scattered trees, and bushes. Rather than fleeing

Table 33. Site SDI-6010 Species List.

Common Name	Scientific Name	Unidentified	NISP*	MNI**
Mammals				
Squirrel	*Amnospermophilus sp.*		3	3
Dog/Coyote	*Canis sp.*		10	4
Kangaroo Rat	*Dipodomys sp.*		2	1
Jackrabbit	*Lepus sp.*		18	7
Wood Rat	*Neotoma fuscipes*		2	2
Deer	*Odocoileus sp.*		1	1
Pocket Mouse	*Pergonathus sp.*		2	2
Raccoon	*Procyon loctor*		1	1
Squirrel	*Sciurus sp.*		3	1
Pocket Gopher	*Thomomys bottae*		48	13
Gopher	*Thomomys sp.*		2	1
Desert Cottontail	*Sylvilagus auduboni*		19	4
Cottontail	*Sylvilagus sp.*		48	8
Guadalupe Fur Seal	*Arctocephalus townsendi*		1	1
	cf. Canis sp.		2	1
	cf. Amnospermophilus sp.		4	1
	cf. Pergonathus sp.		1	1
	cf. Lepus sp.		1	1
	cf. Thomomys sp.		10	11
		very small mammal	5	4
		small mammal	1130	
		medium mammal	46	9
		large mammal	11	2
subtotal			1370	79
Aves (Birds)				
Duck	*Anas sp.*		1	1
Gadwall	*Anas streptera*		1	1
Killdeer	*Charadrius sp.*		1	1
American Coot	*Fulica americana*		1	1
Meadowlark	*Sturnella sp.*		1	1
Morning Dove	*Zenaida sp.*		1	1
Sparrow	*Zonotricha sp.*		1	1
Pigeon	*cf. Columbidae sp.*		1	1
	cf. Zenaida sp.		2	1
		very small bird	9	6
		small bird	2	1
		medium bird	4	3
subtotal			25	19
Elasmobranchs (Sharks and Rays)				
Bat Ray	*Myliobatis californica*		2	1
Shovelnose Guitarfish	*Rhinobatos productus*		2	2
Leopard Shark	*Triakis semifasciata*		2	3
subtotal			6	6
Osteichtyes (Boney Fish)				
Yellowfin Croaker	*Umbrina roucador*		1	1
True Crab	*Brachyura sp.*		1	1
Unidentified Boney Fish	*Unk. Osteichthyes sp.*		5	11
subtotal			7	13
Herpefauna (Reptiles and Amphibians)				
Gopher Snake	*cf. Pitouphus sp.*		2	1
subtotal			2	1
TOTAL			1410	118

*NISP (Number of Identified Specimens)
**MNI (Minimum Number of Individuals)

Table 34. SDI-6010 Frequency and Percentage of all Vertebrates.

Scientific Name	Unidentified	NISP*	%NISP	Grams	% Grams	MNI**	%MNI
Mammals							
Amnospermophilus sp.		3	0.22	1.3	0.81	3	2.54
Canis sp.		10	0.72	5.9	3.68	4	3.39
Dipodomys sp.		2	0.14	0	0.00	1	0.85
Lepus sp.		18	1.30	2.5	1.56	7	5.93
Neotoma fuscipes		2	0.14	0.1	0.06	2	1.69
Odocoileus sp.		1	0.07	0.2	0.12	1	0.85
Pergonathus sp.		2	0.14	0	0.00	2	1.69
Procyon loctor		1	0.07	0	0.00	1	0.85
Sciurus sp.		3	0.22	0.1	0.06	1	0.85
Thomomys bottae		48	3.46	2.5	1.56	13	11.02
Thomomys sp.		2	0.14	0	0.00	1	0.85
Sylvilagus auduboni		19	1.37	1.7	1.06	4	3.39
Sylvilagus sp.		48	3.46	5	3.12	8	6.78
Arctocephalus Townsendi		1	0.07	0	0.00	1	0.85
cf. Canis sp.		2	0.14	0.1	0.06	1	0.85
cf. Amnospermophilus sp.		4	0.28	0	0.00	1	0.85
cf. Pergonathus sp.		1	0.07	0	0.00	1	0.85
cf. Lepus sp.		1	0.07	0.1	0.06	1	0.85
cf. Thomomys sp.		10	0.71	2.1	1.31	11	9.32
	very small mammal	5	0.36	0.1	0.06	4	3.39
	small mammal	1130	81.41	111.9	69.76		0.00
	medium mammal	46	3.31	19.6	12.22	9	7.63
	large mammal	11	0.79	5.5	3.43	2	1.69
subtotal		1370	98.70	158.7	98.93	79	66.95
Aves (Birds)							
Anas sp.		1	0.07	0	0.00	1	0.85
Anas strepera		1	0.07	0.2	0.12	1	0.85
Charadrius sp.		1	0.07	0	0.00	1	0.85
Fulica americana		1	0.07	0.2	0.12	1	0.85
Sturnella sp.		1	0.07	0	0.00	1	0.85
Zenaida sp.		1	0.07	0	0.00	1	0.85
Zonotricha sp.		1	0.07	0	0.00	1	0.85
cf. Columbridae sp.		1	0.07	0.1	0.06	1	0.85
cf. Zenaida sp.		2	0.14	0	0.00	1	0.85
	very small bird	9	0.65	0.01	0.01	6	5.08
	small bird	2	0.14	0.01	0.01	1	0.85
	medium bird	4	0.29	0.01	0.01	3	2.54
subtotal		25	1.80	0.53	0.33	19	16.10
Elasmobranchs (Sharks and Rays)							
Myliobatis californica		2	0.14	0	0.00	1	0.85
Rhinobatos productus		2	0.14	0	0.00	2	1.69
Triakis semifasciata		2	0.14	0.08	0.05	3	2.54
subtotal		6	0.43	0.08	0.05	6	5.08
Osteichtyes (Bony Fish)							
Umbrina roucador		1	0.07	0	0.00	1	0.85
Brachyura sp.		1	0.07	0	0.00	1	0.85
Unk. Osteichthyes sp.		5	0.35	1.1	0.69	11	9.32
subtotal		7	0.50	1.1	0.69	13	11.02
Herpefauna (Reptiles and Amphibians)							
cf. Pitouphus sp.		2	0.14	0	0.00	1	0.85
subtotal		2	0.14	0	0.00	1	0.85
TOTAL		1410	100.00	160.41	100.00	118	100.00

*NISP (Number of Individual Specimens)

**MNI (Minimum Number of Individuals)

from predators, the cottontail runs toward its closest burrow to hide, whereas the jackrabbit utilizes less plant coverage to flee when threatened. *Sylvilagus* prefers to eat leafy green succulents and is known to cause considerable damage to gardens and crops. As with jackrabbits, cottontails were hunted in drives, but due to the secretive nature of the animal and its tendency to hide in burrows when threatened, jackrabbits were easier to obtain by this method. The relative percentages of hares and rabbits indicate that these specimens were commonly hunted for food.

Sylvilagus elements were differentiated from *Lepus* specimens primarily by size as well as cranial osteology.

SCIURIDAE
Amnospermophilus sp.

The members of this genus are small, weighing one to two and one half pounds. They inhabit pastures, grain fields, and slopes with scattered trees and rocky ridges, but avoid thick chaparral and dense woods. This squirrel eats green vegetation, seeds, acorns, mushrooms, fruits, berries, bird eggs, and insects and are considered to be detrimental to crops and pasturelands. The bones of this genus were rarely modified but may have been used for food because they are easily obtained and abundant.

Sciurus sp.

This large squirrel weighs from 1.25 to 1.75 pounds. *Sciurus* prefers fairly open oak and pine forests. This mammal is arboreal, but may often be seen on the ground. The members of this squirrel genus feed on acorns and seeds of conifers. *Sciurus* has been known to be destructive to walnut and almond crops and is considered a fair game animal.

HETEROMYIDAE
Dipodomys sp.

Being the smallest member of the kangaroo rat family, this genus is nocturnal and prefers sandy to rocky soils and mostly low desert with scattered vegetation. This animal feeds mostly on seeds and green vegetation. Because of its small size (1.20 to 1.75 ounces), this rodent is of little economic importance.

CRICETIDAE
Pergonathus sp.

Pocket mice prefer valleys and slopes covered with chaparral or live oaks. These mice are adapted for semi-arid conditions. Their diet consists mostly of seeds. The mice store seeds at their underground nest sites. Their eating of weed seeds is considered beneficial.

Neotoma fuscipes

These wood rats, also known as pack-rats, inhabit rock cliffs and flat areas near washes. Their unique foraging behavior makes their distinct nests easily recognizable. On level areas, these rats build stick houses and are seldom seen by day.

CANIDAE
Canis sp.

These mammals include the domesticated dog and the coyote. The two species inhabit prairies, open woodlands, brushy or boulder strewn areas and usually do not range more than 6 miles from water.

The coyote (*Canis latrans*) scavenges and preys on all types of animals and food plants. Its hunting range extends up to 10 miles from its subterranean den. Coyotes have generally been hunted for their pelts rather than for food.

Due to similarities in social make-up, domesticated dogs (*Canis familiaris*) have accompanied humans for thousands of years. The size of the bones recovered from the prehistoric component at SDI-6010 indicate that the canids weighed between 20 and 50 pounds. Given the age of the site, it is possible that the *Canis* remains could represent either coyote or dog. Without cranial elements, specifically mandibles with ramus, maxillas, or tooth rows, it is extremely difficult to distinguish between the two species.

PROCYONIDAE
Procyon lotor

The raccoon, a nocturnal animal, is approximately the size of a small dog. It inhabits lake borders near wooded areas or rock cliffs along streams. This omnivore eats fruits, nuts, grains, insects, frogs, crayfish, bird eggs, and just about anything else that is available. This animal is potentially destructive to stored foods or even trash and is known to raid or scavenge at human settlements (Gibson et al. 1990). The pelt of the raccoon has economic value and its meat is an important food source.

CERVIDAE
Odocoileus sp.

The members of this genus prefer habitation zones of riparian vegetation and browse shrubs such as coniferous forests, desert shrubs, chaparral, and grasslands with shrubs. The deer is considered one of the most important game mammals of the West due to the size of the animal and because every part of the deer except the gut was either eaten or otherwise utilized.

PHOCIDAE
Arctocephalus townsendi

This Guadualupe fur seal is not a positive identification. The element, a prezygapophys of a thoracic vertebra, resembles that of a black bear but is more similar to the California sea lion. All other pinnipeds have been ruled out and, therefore, this specimen may be the Guadualupe fur seal (a comparative specimen was not available for positive identification). It should also be noted that the fur seal has been the most common pinniped represented in southern California coastal and island prehistoric sites (Bleitz-Sanburg 1987).

Reptiles

Approximately 0.14 percent of the total number of bones belong to one represented reptile. This cold blooded creature accounts for 0.85 percent of the total MNI and less than 0.1 percent of the total bone mass (see Table 34).

COLUBRIDAE
Pituophus sp.

Members of this gopher snake genus average over 1 meter in body length, have a stout build, and are terrestrial, non-venomous, and powerful constrictors. This snake genus ranges across the continental United States, north to south Alberta and Saskatchewan, Canada and to south central Mexico. They are especially abundant in grasslands, open woodlands and can be found over all areas above the tidal flux. Adults are diurnally active through July and have peak surface activity in May.

Fish

The Elasmobranchs account for 0.43 percent of the total number of bones and Osteichthyes 0.50 percent. Their total bone masses are 0.05 percent and 0.69 percent, respectively. The Osteichthyes have a total MNI of 11.02 percent of all fauna, whereas the Elasmobranchs account for 5.08 percent (see Table 34).

TRIAKIDAE
Triakis semifasciata

The Leopard shark ranges in length from 2 to 7 feet. It is found in all seas and occasionally in brackish or fresh water. This species can be found on sand and mud and inhabits mainly inshore and shallow bays (Salls 1988).

RHINOBATIDAE
Rhinobatus productus

The shovelnose guitarfish are strange-looking rays and are 1 to 5.5 feet in length. Its habitat is sand or mud bottoms in shallow seas and brackish waters. They swim close to the bottom hunting small shellfish.

MYLIOBATIDAE
Myliobatis californica

This bat ray can weigh up to 210 lbs. and have a width of 6 feet. It ranges from Oregon to the Gulf of California. The bat ray prefers bays and other shallow, sandy or mud bottoms to depths of 150 feet.

SCIAENIDAE
Umbrina roncador

The yellowfin croaker is a bony fish and can range in length up to 20 inches. It prefers habitats around kelp beds and over soft bottoms from surf zones to depths of 150 feet. It ranges from Point Conception, California to the Gulf of California.

DROMIIDAE
Brachyura sp.

A crab claw fragment was recovered from the faunal assemblage. The size of the crab is not known. Crabs can be found from British Columbia to Baja California in tide pools and deeper sandy bottoms.

Birds

The Aves fauna are the second most abundant and form 16.10 percent of the total MNI. Their total bone mass is 0.33 percent and the number of Aves bones account for 1.80 percent of the total (see Table 34).

ANATIDAE
Anas sp.

The members of this duck genus prefer marshes, ponds, lakes, rivers, bays, and flooded fields. They feed in deeper open water, yet they also graze on marshes and meadows. They eat wetland plants and seeds, small aquatic animals, fish, and molluscs. Ducks are mostly migratory.

Anas stereptera

The gadwall inhabits marshes, ponds, and rivers. It winters in the West where waters do not freeze, in south-central and eastern seaboard states. Gadwalls are dabbling ducks, in that they feed on the bottom in shallow water by tipping forward allowing the tail to stick up.

ICTERIDAE
Sturnella sp.

Also known as the meadowlark, this robin-sized passerine prefers grassy fields, and its diet consists of insects, and fruits (even nectar or seed). It winters north along the Pacific coast from Mexico to southwest British Columbia, inland to Idaho, and Wyoming to Nebraska.

COLUMBIDAE
Zenaida sp.

The mourning dove ranges from sparrow to duck size and prefers open country. They feed on fruits, seeds, and foliage, and they range from south Canada to the Mexican border. They winter throughout their range except Canada and the northern border states. Breeding season starts in March (in California) and continues to mid-September. The dove may nest two to four times each year, thus making it easily able to maintain its numbers even though it is hunted extensively.

RALLIDAE
Fulica americana

The American coot is chicken-sized and inhabits ponds, bays, lakes, and marshes, in which they skulk in dense reed beds. They are migratory and winter where waters do not freeze and can be found even on salt water if submerged vegetation on which to feed is available, such as aquatic vegetable or animal matter.

CHARADRIIDAE
Charadrius sp.

The members of the killdeer genus are sparrow- to pigeon-sized "waders" and frequent shores of lakes, rivers, and the edges of marshes, as well as tidal flats and even ocean beaches. They eat snails and other types of animal life directly from the ground or from shallow pools, mud flats, and beaches. The killdeer are migratory and winter on the coast to north California, south Nevada, Arizona, New Mexico, Oklahoma and southeast Kansas. They feed on animal life in wetlands, running about on beaches, marshes, and meadows.

FRINGILLIDAE
Zonotrichia sp.

Members of this genus are called sparrows and prefer habitats such as coniferous forests, woodlands, and openings. The ranges of this family are widespread through the United States, Canada, and the northern parts of Mexico. The sparrow winters mostly along the west coast.

Unidentifiable Bone

The ability to identify bone is largely dependent upon, but not limited to, the degree of fragmentation. The density of the bone, age of the site, trampling, and the amount of overburden are all factors in bone identifiability. The percentage of unidentifiable bone at the prehistoric component at SDI-6010 is 85 percent. Although relatively high, this percentage is not unusual when compared to other prehistoric sites, and is probably due to the fact that the aboriginal inhabitants of southern California usually pounded an entire animal in a grinding dish then mixed it in with vegetable foods for consumption (Kroeber 1925). Of the total of 1,410 bones recovered from the site, 1,193 fragments could not be identified.

EXPLOITED HABITATS AND HUNTING PRACTICES

The animals represented in the SDI-6010 faunal assemblage lived in a variety of ecozones, ranging from near shore aquatic environments, grasslands, and chaparral to low valleys with grass, sagebrush, scattered trees, and bushes. The fish recovered all inhabit lagoon or near shore environments. There is no evidence of deep-sea fishing. With the exception of the fish, which would have been restricted to the lagoon 5 km downstream, all animals represented in the faunal collection would have been available near the site.

Although no evidence of traps, snares or nets was recovered, it is reasonable to assume that the inhabitants of the site used a variety of specialized hunting strategies. As mentioned earlier, rabbit drives to obtain hares and cottontails are well-documented for the ethnohistoric period. The rabbits were driven into long, low nets where rabbit sticks and clubs were used to dispatch the animals (Bean and Shipek 1978). Jackrabbits are the easier of the two to capture due to their nature of fleeing when threatened rather than hiding in the nearest burrow as do cottontails. For this reason, cottontails are usually obtained by solitary hunters. The fact that cottontails outnumber jackrabbits by a margin of about two to one at the Whelan Lake site, suggests that solitary hunting methods were preferred.

These natives were also known to be experts in stalking animals with the bow and arrow. This technique allowed them to capture coyotes, deer, rabbits, raccoons, and other fauna.

Another hunting strategy that may have been utilized by the residents of the SDI-6010 is the use of deadfalls. These are traps set for small rodents where two stones were used. The upper stone was propped up with a light stick resting on some type of bait. When the animal gnawed through the bait the stone would fall and kill the animal. Deadfalls were commonly used by many California Indians during the ethnohistoric period when hunting small rodents (Kroeber 1925).

Small fish may have been caught on hook and line using live bait or in nets along the shore or in the lagoon. In addition, spearing may have been utilized in shallow water. These methods may have also been used for the leopard shark.

Determining seasonality is difficult with the current faunal assemblage from the site. All bird species recovered from the site inhabit the Pacific coast year-round. It may be of interest to note that remains of migratory waterfowl, which are currently present in the area between late fall and early spring, were not recovered. Their absence could be interpreted in three ways. First, these animals were hunted and their remains were deposited on site, but the current sample was not of sufficient size to recover them. Second, the site was occupied during the fall, winter, and spring, but these birds were not hunted. Third, the site was not occupied during this part of the year.

Given the small number of identifiable bird bone, it is quite likely that migratory waterfowl were hunted and not identified in the assemblage. In contrast, it seems unreasonable that if these species were available that the residents of the site would not have hunted them. These conclusions still leave the issue of seasonality in doubt. Only the analysis of additional faunal assemblages from this time period can resolve this issue.

Animal Processing

Of the total number of bones, 51 (14%) were burned (Table 35). Of these, particular interest was given to the cottontail and hare specimens. Most of these specimens represented phalanges, maxillae, ischia, and metacarpals: elements that are only covered by thin tissue and muscle. These observations are consistent with processing practices in which the animals were cooked over an open fire where the elements were exposed to a flame. Some elements were also extremely calcined, a result that could only be due to extreme heat, such as in the ashes of a hearth or a roasting pit.

The high percentage of fragmented bones is consistent with assemblages from throughout southern California. On the basis of ethnographic data, Shelley (in Altschul and Shelley 1987) has argued that a fragmented assemblage is consistent with the use of bone as an additional source of fat. For example, the Nunamuit Eskimo put bone in stews and other foods, which were then boiled. This process extracted grease and fat from the bone, thereby providing important dietary supplements in times of stress.

This source of fat was especially important near the end of winter and during the early spring when available fresh animal meat tended to be very lean. Among the Nunamuit, bone juice processing occurred whenever it was not possible to store the bones for gang processing, primarily due to high temperatures spoiling the fats, blood, and marrow. Under these conditions, the Nunamuit often processed a heterogeneous mixture of bone by first crushing it, then adding it to various cooking foods to produce a juicy broth that had grease and fats. The result of this processing is numerous small chips and fragments of bones, generally of long bones, phalanges, ribs, and mandibles. At Whelan Lake the faunal assemblage was dominated by small fragments of long bones, a composition consistent with the use of bone for obtaining fat.

Bones that had been butchered or polished were also found at SDI-6010. Cut marks were found on 12 bones. Of these, one *Sylvilagus* bone shows a cut mark which runs in a transverse orientation to the length of the bone, suggesting the skinning of these animals (Figure 34a). Ethnographically, such pelts of fur-bearing animals were highly prized (Heizer and Elsassen 1980:157; Kroeber 1925:805).

One fragmented piece of worked bone was also found. It shows both cut marks and has a polished finish to its surface (Figure 34b). The piece was too small to determine its function.

Table 35. Burned Bones Recovered from CA-SDI-6010.

Level	Feature	Test Pit	Taxon	Element	Number of Fragments	Percentage of Fragments
3		1	Small Mammal	longbone fragment	1	0.53
3		1	Small Mammal	longbone fragment	1	0.53
4		1	*Neotoma fuscipes*	proximal tibia	1	0.53
4		1	Small Mammal	indeterminate	2	1.05
1	1*	2	Small Mammal	indeterminate	4	2.11
2	1*	2	Small Mammal	indeterminate	28	14.74
3	1	2	Small Mammal	longbone fragments	8	4.21
3	1	2	*Sylvilagus sp.*	radius	1	0.53
3	1	2	*Sylvilagus sp.*	rib fragment	1	0.53
3	1	2	*cf. Canis sp.*	proximal rib	1	0.53
1		4	Medium Mammal	longbone fragment	1	0.53
5		4	*Lepus sp.*	calcaneum	1	0.53
1		6	Small Mammal	longbone fragment	1	0.53
2		6	Small Bird	longbone fragments	2	1.05
4		6	Small Mammal	longbone fragments	2	1.05
4		6	Very Small Mammal	indeterminate	1	0.53
5		6	Small Mammal	indeterminate	1	0.53
6		6	Small Mammal	indeterminate	2	1.05
7		6	Small Mammal	indeterminate	1	0.53
1		7	Medium Mammal	indeterminate	2	1.05
3		7	Medium Mammal	indeterminate	1	0.53
3		7	Small Mammal	longbone fragment	1	0.53
1		8	Small Mammal	indeterminate	13	6.84
1		8	*Sylvilagus sp.*	metatarsal	2	1.05
2		8	Medium Mammal	indeterminate	3	1.58
2		8	Small Mammal	indeterminate	33	17.37
2		8	*Sylvilagus auduboni*	phalanx	1	0.53
2		8	*Sylvilagus sp.*	proximal radius	1	0.53
3		8	Medium Mammal	longbone fragments	7	3.68
3		8	Small Mammal	longbone fragments	14	7.37
3		8	*Sylvilagus auduboni*	ischium	1	0.53
3		8	*Sylvilagus sp.*	metatarsal	1	0.53
4		8	Small Mammal	longbone fragments	9	4.74
5		8	Large Mammal	longbone fragment	1	0.53
5		8	Small Mammal	longbone fragments	5	2.63
7		8	Medium Mammal	indeterminate	1	0.53
7		8	Small Mammal	longbone fragments	5	2.63
1		9	Very Small Mammal	longbone fragment	1	0.53
2		9	Small Mammal	longbone fragment	1	0.53
2		9	Small Mammal	indeterminate	1	0.53
4		9	Small Mammal	longbone fragments	3	1.58
2		10	Small Mammal	longbone fragment	1	0.53
2		10	Small Mammal	longbone fragment	1	0.53
3		10	Medium Mammal	indeterminate	2	1.05
3		10	Small Mammal	longbone fragment	1	0.53
4		10	Small Mammal	longbone fragments	6	3.16
5		10	*Sylvilagus sp.*	incisor	1	0.53
2		11	Medium Mammal	longbone fragments	2	1.05
2		11	Small Mammal	longbone fragments	3	1.58
1		12	Small Mammal	longbone fragments	5	2.63
2	2	12	*Sylvilagus auduboni*	maxillary process	1	0.53
TOTAL					190	100

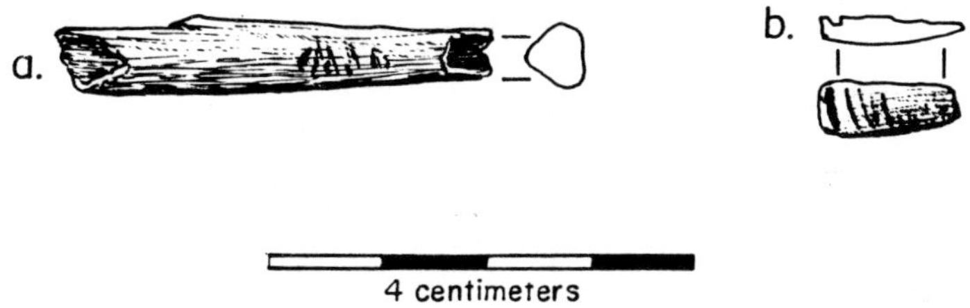

Figure 34. Worked bones from SDI-6010: (a) *Sylvilagus* bone showing cut marks running transverse to length, (b) fragment of worked bone.

Intrasite Analysis

Table 36 presents the NISP for all test pits and levels. When examining Table 36 it is important to remember that Test Pits 2 and 8 were 2 m x 2 m units, whereas the others were all 1 m x 1 m units. NISP calculations for the two former test pits, then, should be divided by four before comparing them with other units.

Faunal remains were distributed roughly in relation to their position in the midden. Units closest to the center of the site, where the midden was thickest, contained more bones in greater densities than those further away. Bone frequencies ranged from zero for Test Pit 3, located in a field west of the site, to 966 for Test Pit 8, located in the exact center of the site. Even if one divides this amount by four, Test Pit 8 still contained more than three times the number of faunal remains of any other unit.

The vertical distribution of faunal remains is intriguing. Table 36 shows that peaks in NISP vary across the site. Test Pit 8, in the center of the site, has a unimodal distribution of faunal remains, showing a sharp peak in Level 2 and to a lesser extent in Level 3. Test pits in the southwest portion of the site (i.e., Test Pits 1, 6, 7, and 9) also are unimodal in their distribution of faunal elements, but the peak tends to be lower in the midden, between Levels 4 and 5. These shifts may reflect changes in activity areas over time, with the center of the site moving from the southwest quadrant to a more central location late in site's occupational history.

Regardless of changes in activity areas, little change can be discerned in the nature of animal species exploited. Throughout the site's history, emphasis was placed on procuring small terrestrial mammals, primarily hares and rabbits.

INTRASITE ANALYSIS AND LA JOLLAN ADAPTATION

The Whelan Lake site is the earliest dated component in the lower portion of the San Luis Rey River basin. Sites of similar age have been found in nearby drainages, notably the Windsong Shores (SDI-10965) and Allan O. Kelly (SDI-9649) sites in the Aqua Hedionda lagoon just south of San Luis Rey (Gallegos 1991), the Rancho Park North sites about 15 km south of Whelan Lake (Kaldenberg 1982), and the San Dieguito Estates about 25 km south of SDI-6010 (Norwood and Walker 1980). The Windsong Shores site has two components, the earliest dating between 8,400 and 8,000 B.P. and the later around 7,000 B.P., whereas the Allan O. Kelly site has a continuous occupation between 7,500 to 6,850 B.P. The sites at Rancho Park North also exhibit two periods of occupation, the earlier between 8,300 and 8,000 B.P. and the later again around 7,000 B.P. The San Dieguito Estates sites are more continuous in their occupation, dating between 8,650 and 7,400 B.P.

Table 36. NISP by Provenience.

Legend			
1	*Amnospermophilus sp.*	22	Medium Mammal
2	*Canis sp.*	23	Large Mammal
3	*Dipodomys sp.*	24	*Anas sp.*
4	*Lepus sp.*	25	*Anas streptera*
5	*Neotoma fucipes*	26	*Charadrius sp.*
6	*Odocoileus sp.*	27	*Fulica americana*
7	*Pergonathus sp.*	28	*Sturnella sp.*
8	*Procyon loctor*	29	*Zenaida sp.*
9	*Sciurus sp.*	30	*Zonotricha sp.*
10	*Thomomys bottae*	31	cf. Columbidae sp.
11	*Thomomys sp.*	32	*cf. Zenaida sp.*
12	*Sylvilagus auduboni*	33	Very Small Bird
13	*Sylvilagus sp.*	34	Small Bird
14	*Arctocephalales towsendi*	35	Medium Bird
15	*cf. Canis sp.*	36	*Myliobatis californica*
16	*cf. Amnospermophilus sp.*	37	*Rhinobutus ptoductus*
17	*cf. Pergonathus sp.*	38	*Triakis semifasciata*
18	*cf. Lepus sp.*	39	*Umbrina roucador*
19	*cf. Thomomys sp.*	40	*Brachyura sp.*
20	Very Small Mammal	41	Unk. Osteichthyes sp.
21	Small Mammal	42	*cf. Pitouphis sp.*

Table 36. NISP by Provenience (Continued)

	Species													
	1	2	3	4	5	6	7	8	9	10	11	12	13	14
Test Pit 1														
Level 1													1	
Level 2														
Level 3														
Level 4					1									
Level 5														
Test Pit 2														
Level 1										1			1	
Level 2				2						1			1	
Level 3		2					1						7	
Level 4														
Test Pit 4														
Level 1														
Level 2														
Level 4										1				
Level 5				1										
Test Pit 5														
Level 3										1				
Level 6										1				
Level 8														
Test Pit 6														
Level 1														
Level 2	1					1								
Level 3														
Level 4														
Level 5				1						1		1	1	
Level 6														
Level 7										1				
Test Pit 7														
Level 1														
Level 3														
Level 4										1				
Level 5										1				
Level 6														
Test Pit 8														
Level 1		1								2			4	
Level 2		1	2	3	1				2	10		8	9	
Level 3		2		5				1	1	7		5	4	
Level 4	1						1			4		2	2	1
Level 5	1	1		1						3			3	
Level 6				1						4		1		
Level 7										1			1	
Test Pit 9														
Level 1														
Level 2		1												
Level 4				1										
Level 6										1				
Level 7										1				
Test Pit 10														
Level 1										1				
Level 2		2					1							
Level 3										1			2	
Level 4		1										1	3	
Level 5				1						1			2	
Test Pit 11														
Level 1										1			1	
Level 2														
Level 3														
Test Pit 12														
Level 1				1						1			2	
Level 2				1						1	2	1	4	
Total	3	11	2	18	2	1	3	1	3	48	2	19	48	1
% Total	0.21	0.78	0.14	1.28	0.14	0.07	0.21	0.07	0.21	3.40	0.14	1.35	3.40	0.07

Table 36. NISP by Provenience (Continued)

	Species													
	15	16	17	18	19	20	21	22	23	24	25	26	27	28
Test Pit 1														
Level 1							1							
Level 2							1							
Level 3							4							
Level 4							11							
Level 5							2							
Test Pit 2														
Level 1							19						1	
Level 2							28	5						
Level 3							63	3						
Level 4														
Test Pit 4														
Level 1								1						
Level 2							1							
Level 4							1							
Level 5														
Test Pit 5														
Level 3														
Level 6								1						
Level 8							2							
Test Pit 6														
Level 1							1							
Level 2							1		1					
Level 3							1		1					
Level 4						1	3		2	1				
Level 5							13							
Level 6							2							
Level 7							2							
Test Pit 7														
Level 1						1		2						
Level 3							13	1						
Level 4		1					5				1			
Level 5							1							
Level 6							1							
Test Pit 8														
Level 1							73							
Level 2		2			3	1	373	3	2					
Level 3		1			4		233	7	1			1		1
Level 4					3		49	3	1					
Level 5	1						42	4	3					
Level 6							24							
Level 7				1			15	1						
Test Pit 9														
Level 1						1								
Level 2						1	8							
Level 4							15	1						
Level 6							3							
Level 7							3							
Test Pit 10														
Level 1							6							
Level 2							12	1						
Level 3							16	5						
Level 4							9	3						
Level 5							9	1						
Test Pit 11														
Level 1							11							
Level 2							15	3						
Level 3							12							
Test Pit 12														
Level 1							15							
Level 2							11	1						
Total	1	4	0	1	10	5	1130	46	11	1	1	1	1	1
% Total	0.07	0.28	0.00	0.07	0.71	0.35	80.14	3.26	0.78	0.07	0.07	0.07	0.07	0.07

Table 36. NISP by Provenience (Continued)

Species	29	30	31	32	33	34	35	36	37	38	39	40	41	42	Total	% Total
Test Pit 1																
Level 1															2	0.14
Level 2															1	0.07
Level 3					1										5	0.35
Level 4				1					1						14	0.99
Level 5															2	0.14
Test Pit 2																
Level 1															22	1.56
Level 2															37	2.62
Level 3															76	5.39
Level 4							1								1	0.07
Test Pit 4																
Level 1															1	0.07
Level 2															1	0.07
Level 4															2	0.14
Level 5															1	0.07
Test Pit 5																
Level 3															1	0.07
Level 6															2	0.14
Level 8															2	0.14
Test Pit 6																
Level 1															1	0.07
Level 2						2									6	0.43
Level 3							2								4	0.28
Level 4															7	0.50
Level 5					1										18	1.28
Level 6															2	0.14
Level 7															3	0.21
Test Pit 7																
Level 1					1										4	0.28
Level 3					1										15	1.06
Level 4					1										9	0.64
Level 5															2	0.14
Level 6															1	0.07
Test Pit 8																
Level 1													1	2	83	5.89
Level 2	1	1	1	1			1			1	1		1		428	30.35
Level 3												1	2		276	19.57
Level 4								1							68	4.82
Level 5					1			1							61	4.33
Level 6									1						31	2.20
Level 7															19	1.35
Test Pit 9																
Level 1													1		2	0.14
Level 2															10	0.71
Level 4															17	1.21
Level 6															4	0.28
Level 7															4	0.28
Test Pit 10																
Level 1															7	0.50
Level 2					1										17	1.21
Level 3															24	1.70
Level 4										1					18	1.28
Level 5															14	0.99
Test Pit 11																
Level 1					2										15	1.06
Level 2															18	1.28
Level 3															12	0.85
Test Pit 12																
Level 1															19	1.35
Level 2															21	1.49
Total	1	1	1	2	9	2	4	2	2	2	1	1	5	2	1410	100.00
% Total	0.07	0.07	0.07	0.14	0.64	0.14	0.28	0.14	0.14	0.14	0.07	0.07	0.35	0.14	100.00	-

The Rancho Park North and San Dieguito Estates sites are similar to Whelan Lake in their setting. All three locations are between 5 and 7.5 km from the nearest downstream lagoon. Each area produced sites with faunal assemblages of similar composition. Fish, bird, and small mammal remains characterize these collections, indicating a widespread emphasis on animals from a diversity of ecological settings.

Unlike Whelan Lake, the Aqua Hedionda sites were situated near the edge of an ancient lagoon. As might be expected given their proximity to the lagoon, their assemblages contain greater amounts and diversity of fish and aquatic bird remains. Still, the assemblages from Aqua Hedionda and Whelan Lake are quite similar, with the vast preponderance of the collection being small mammal, particularly rabbit and rodent.

Within the San Luis Rey River drainage, the site most similar to Whelan Lake in terms of geographic location and date of occupation is SDI-5130 and its associated site, SDI-6014 (Gibson et al. 1990). SDI-5130 is a large, habitation site whose main component dates between 6,500 and 4,000 B.P., with a much smaller occupation occurring after A.D. 1400. It is located on a terrace overlooking the north bank of the San Luis Rey River about 1.5 km downstream from SDI-6010.

The faunal assemblages of SDI-5130 and SDI-6010 are extremely similar (Table 37). Both are dominated by small mammals, particularly rabbit and rodent. Terrestrial mammals account for about 67 percent of the SDI-6010 collection as opposed to 78 percent at SDI-5130. These percentages inflate the importance of mammals because both include gopher and other rodent bones that are probably intrusive. Moreover, the SDI-5130 assemblage includes the remains of a number of domesticated animals associated with an historic occupation of the site. Birds and fish were found in somewhat greater numbers at SDI-6010, whereas reptiles were in less quantities. A chi-square test of the MNI distribution between the four classes of animals indicates that the observed distribution is unlikely to be due to chance alone. Over time, then, it appears that there was an increasing focus on the hunting of mammals, with less attention placed on aquatic birds and lagoonal fish.

Table 37. Comparison of MNI in the Faunal Collections from SDI-5130 and SDI-6010.

	SDI-5130		SDI-6010	
	N	%	N	%
Terrestrial Mammal	250	77.88	90	66.95
Bird	20	6.23	19	16.10
Fish	40	12.43	19	16.10
Reptiles	11	3.46	1	0.85

Chi-square = 11.15
degrees of freedom = 3
alpha level > .01

This finding is surprising because SDI-5130 would have been located closer to the salt water marsh and lagoon than SDI-6010. Although enigmatic, these results parallel those of the invertebrate faunal analysis, which indicate that a greater emphasis was placed on marine mollusc at SDI-6010 than at SDI-5130.

Explaining these observations is difficult. It is possible that the geomorphic interpretation is in error, and that the lagoon extended all the way to Whelan Lake. Orme (personal communication 1991) recognizes this possibility, but considers it very unlikely. He believes that any salt water intrusion as far upstream as Whelan Lake would have been rather short-lived. The mouth of the river would have been very close by, transforming salt water marshes to fresh water ones in a relatively short period of time.

The present evidence suggests that the area around the site consisted of a fresh or brackish water marsh. With the exception of the fish, the fauna represented at the site could easily have been recovered in the immediate area. Fresh water marshes are highly productive ecological zones, and often were favored locales of early Holocene hunters and gatherers. Gallegos (1991) argues that the early inhabitants of the San Diego region may have originated near similar marshes in the Great Basin and desert regions to the west. If so, the newcomers to the region may have found settling familiar ecological zones a useful first step in developing a coastal adaptation.

CHAPTER 10

MICRO- AND MACROBOTANICAL ANALYSIS

The paleobotanical analysis for CA-SDI-6010 involved the examination of selected pollen and flotation samples collected from excavation units. Main goals for the analyses were to answer questions about subsistence, seasonality, and paleoenvironment. Pollen analysis was conducted by Owen K. Davis, Department of Geosciences, University of Arizona. Flotation samples were analyzed by Charles H. Miksicek from BioSystems, Inc.

POLLEN ANALYSIS

Owen K. Davis

Introduction

Prior to settlement, the vegetation in the area around the Whelan Lake site probably consisted of coastal scrub, oak woodland, and chaparral (Barbour and Major 1988; Affinis 1991). Three wetland vegetation types are recognized at Whelan Lake and the San Luis Rey River near CA-SDI-6010 (Affinis 1991). The Freshwater Marsh contains grasses (Gramineae), sedges (*Cyperaceae*) and cattail (*Typha* spp.), the Alkaline Marsh includes pickleweed (*Salicornia*), and the Riparian Woodland is dominated by willows (*Salix* spp.) and cottonwood (*P. fremontii*). Weeds and introduced plants are abundant, including *Brassica* sp., *Erodium* sp., *Nicotiana glauca*, *Salsola australis*, and *Xanthium spinosum*.

Seven pollen samples were analyzed (Table 38). One of these, Item # 25, is a modern analog sample collected from the A-horizon in the profile of Trench V, located about 250 m northwest of the site. Two samples originate from Feature 1, a hearth initially exposed in Trench G and then excavated with Test pit 2. Of these, Item # 96 came from the trench profile and Item # 272 was taken from beneath a rock near the bottom of the feature. Item # 295 was collected from the fill of the second hearth, Feature 2. The remaining two samples were taken from the general midden fill; Item # 113 is from an area with a high shell density and Item # 125 is from an area with high artifact density. The samples were collected to help address two basic research domains related to the paleoenvironment of the Whelan Lake site:

(1) What was the vegetation at the time of occupation, was the Whelan Lake basin a marsh or lake, how close were aquatic resources to the site?

(2) What was the diet of the occupants of CA-SDI-6010, does the pollen provide ethnobotanical information?

Table 38. Analyzed Pollen Samples from the Whelan Lake Site.

Provenience Number	Item Number	Unit/TP Number	Feature Number	Level	Depth (cm BS)	Comments
30.02	96	Trench G	1	-	60-65	Taken from profile of trench G- before excavation of TP2 (=from ashy bottom fill of hearth)
75	272	2	1	-	65-70	From under rock or bottom of feature
40.02	113	6	-	-	40-45	From area with high shell density
41.02	125	7	-	-	35-40	From area with moist/soft fill and high artifact density
80.02	268	8	-	-	45-50	-
84.03	295	12	2	-	35-40	From fill of pit feature
13.02	25	Trench V	-	-	0-5	Modern analogue sample: from A-horizon in trench profile

Methods

The sediment samples were treated using acid digestion to remove inorganic matter (salts, carbonates, silicates) and easily-dissolved organic matter (Table 39). This technique is preferable to flotation if the pollen concentration is greater than 100 grains/cm^3 because flotation does not completely remove pollen from the inorganic matrix and because acid digestion is less hazardous -- flotation liquids are carcinogenic and environmental pollutants. Tracers were added to determine the concentration of pollen in the samples. Calculation of concentration is useful to detect loss of pollen through deterioration in cases where deteriorated percentage and concentration are negatively correlated. Alternately, if preservation is good and vegetation change is small, the concentration can reflect changes in the rate of sediment accumulation, low concentration rates implying rapid accumulation of the sediment matrix.

A minimum of 300 pollen grains were counted per sample, not including tracers, spores, or charcoal. This provides a statistically reliable representation of rare pollen types, which are useful as indicators.

Results

Pollen preservation was poor to fair (38% - 16% deteriorated) and the concentration was average for archaeological samples (1161-8466 grains/cm^3, Table 40). However, concentration and deteriorated percentages are not correlated, so not much of the original pollen has been lost to deterioration.

Comparison of Archaeological Samples with the Modern Analog

The modern analog sample (Item #25) has a higher concentration, less deteriorated pollen, and higher weed pollen percentages than the archaeological samples. The weed pollen types include Chenopodiaceae-*Amaranthus*, Gramineae (grass), and Cruciferae (mustards). Each of these groups includes numerous introduced weeds in the modern upland vegetation of the Whelan Lake area (Affinis 1991). Cruciferae percentages also are elevated during early settlement or immediate

Table 39. Pollen Extraction Procedure.

Step	
1.	add 20 cm^3 sample of sediment to ca. 50 ml 10% HCl
2.	add 1 tracer tablet (13,500 Lycopodium spores per tablet)
3.	swirl solution and screen (180 micron stainless steel) into second beaker
4.	transfer screened solution to 50 ml Nalgene test tubes, repeat "step 4," centrifuging, until complete
5.	add 10 ml conc. HCl, mix, add 30 ml H_2O, mix, centrifuge, decant, water rinse
6.	add 40 ml HF overnight or 1 hr in boiling water bath, centrifuge, decant, water rinse
7.	add 10 ml conc. HCl, mix, add 30 ml H_2O, mix, centrifuge, decant, water rinse
8.	add 30 ml 10% KOH 2 min. boiling water bath, centrifuge, decant, hot water rinses until clear, transfer to 15 ml glass tubes
9.	acetolysis add 5 ml glacial acetic acid, centrifuge and decant, add 5 ml acetic anhydride (volumetric dispenser), add 0.45 ml H_2SO_4, mix, centrifuge, decant into glacial acetic acid, add 5 ml glacial acetic acid, centrifuge and decant
10.	add 10 ml 10% KOH, 2 min. boiling water bath, centrifuge, decant, hot water rinses until clear
11.	stain with safranin "O"
12.	examine samples microscopically for samples with abundant $<5\ \mu$ debris, Nitex 10 μ screen
13.	add a few drops of glycerin
14.	transfer to labeled 1 dram shell vials

pre-settlement times at the Ballona Lagoon (CA-LAN-60) and San Joaquin Marsh (Davis 1989, 1991) (Figure 35). The modern analog sample contains lower *Artemisia* percentages than the archaeological samples, probably reflecting the reduction of California sagebrush (*Artemisia californica*) following settlement.

The archaeological samples contain high percentages Liguliflorae pollen, whereas the modern analog has none. Liguliflorae (a subfamily within Compositae or Asteraceae) pollen is also abundant (<20%) in archaeological samples from CA-LAN-60 and from 10 sites near San Joaquin Marsh. It is less abundant (<5%) in archaeological sites from the interior (Davis and Miksicek 1988a,b,c). Liguliflorae pollen is rare throughout the 7,000 year record at San Joaquin Marsh, even though archaeological sites surround the marsh. Liguliflorae is not an airborne pollen type. Therefore, its widespread abundance in archaeological sites of southern California must indicate local disturbance conditions.

The plants responsible for elevated Liguliflorae percentages may have been native weeds that were replaced in abundance by the many introduced weeds of the California flora. Although none of the upland species listed near Whelan Lake are in the Liguliflorae subfamily (Affinis 1991), weedy Liguliflorae are abundant in the modern vegetation of coastal California (Schreiber 1981) and their

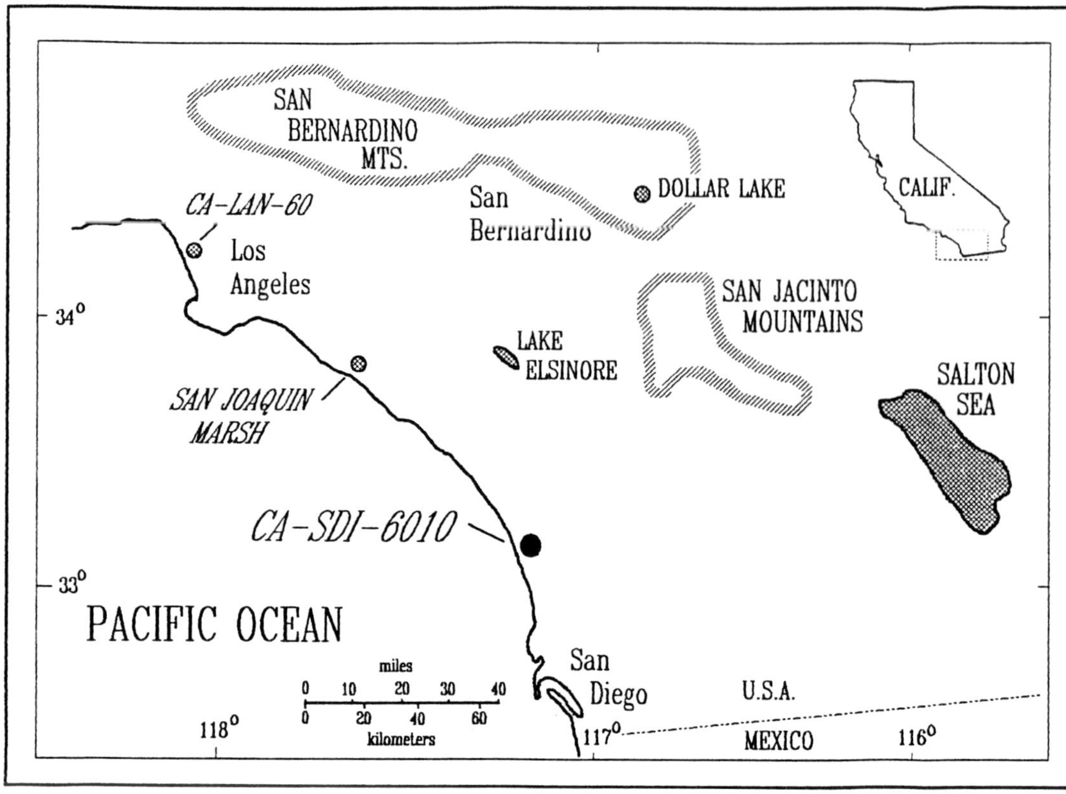

Figure 35. Map of southern California showing the Whelan Lake site and other sites discussed in the text.

pollen is abundant in archaeological samples at CA-LAN-60 (Davis 1991). The hypothetical formerly-abundant weed could be any member of the *Agoseris*, *Malacothrix*, *Microseris*, or *Stephanomeria* genera of the Liguliflorae, abundant in disturbed coastal vegetation. For example, *Stephanomeria virgata* is a common, widespread, late-summer annual of disturbed areas from coastal to interior western North America.

Alternatively, the elevated Liguliflorae percentages might indicate human utilization. None of the ethnobotanically-useful species listed for the San Luis Rey area by Quillen et al. (1984), however, are in the Liguliflorae subfamily.

Erodium cicutarium (storksbill), present at 3.7% (Table 40 and Figure 36) in the modern analog, is one of the most consistent indicators of settlement throughout western North America, and is one of the earliest indicators of European contact at San Joaquin Marsh (Davis 1989). The absence of this indicator, and the lower percentages of Cruciferae and other weedy types confirms that the archaeological samples date to pre-settlement times.

Table 40. Pollen Percentages for CA-SDI-6010.

	Modern Analog	Feature 1 Hearth	Feature 1 Rock	Shell Conc.	Arti-facts	Feature 2
U of A Lab #	1	2	5	3	4	6
SRI Item #	25	96	272	113	125	295
Sum	301	301	306	303	301	303
Concentr. gr/cc	8466	1161	2909	4352	8466	5681
Tracers	24	75	71	47	24	36
Deteriorated	7.0	34.9	16.3	37.6	33.6	17.5
Terrestrial Plants:						
Cupressaceae	0.0	0.0	0.3	0.0	0.0	0.0
Pinus	0.3	0.0	0.0	0.0	0.0	0.0
Quercus	1.0	0.3	0.0	0.3	0.0	0.7
Ceanothus	0.3	0.3	0.0	0.0	0.0	0.0
Ephedra	0.3	0.0	0.0	0.0	0.0	0.0
Cheno-*Amaranthus*	3.7	1.0	2.6	0.7	3.7	5.0
Sarcobatus	0.0	0.0	0.0	0.3	0.0	0.0
Gramineae	6.6	0.7	0.7	0.3	0.0	1.0
Ambrosia	1.0	0.0	0.7	2.0	1.0	0.0
Artemisia	1.7	1.3	4.9	4.3	4.3	8.9
Liguliflorae	0.0	17.6	29.4	11.9	9.6	5.9
Other Compos.	57.1	42.2	44.1	40.3	42.9	56.8
Cruciferae	15.3	0.7	0.3	0.3	0.0	1.0
Eriogonum	0.3	0.3	0.3	1.0	4.0	0.3
Erodium cicutarium	3.7	0.0	0.0	0.0	0.0	0.0
Erodium < > cicutar.	0.3	0.3	0.0	0.0	0.0	0.3
Liliaceae	0.0	0.0	0.3	0.0	0.0	0.0
Onagraceae	0.0	0.0	0.0	1.0	0.3	0.0
Opuntia	0.0	0.0	0.0	0.0	0.0	1.3
Rhus thick striate	0.7	0.3	0.0	0.0	0.7	1.3
Umbelliferae	0.7	0.0	0.0	0.0	0.0	0.0
Aquatic Plants (Not in Pollen Sum):						
Cyperaceae	1.7	0.3	1.3	0.0	0.3	0.0
Potamogeton	0.3	0.0	0.0	0.0	0.0	0.0
Typha latifolia	0.0	0.0	0.0	0.3	0.0	0.3
Typha-Sparganium	0.0	0.0	0.0	0.7	0.0	0.7
Spores and Microfossils (Not in Pollen Sum):						
Fern Spores	2.7	0.0	0.0	0.0	0.0	0.0
Sporormiella	1.7	0.0	0.0	0.7	0.0	0.0
Thecaphora	0.0	0.0	0.7	1.0	3.0	3.3
Other Fungal	185.4	69.8	35.3	69.3	138.2	39.3
Charcoal	804.0	1189.7	91.5	462.0	725.6	807.6
Fecal Pellets	1.0	8.0	4.9	6.3	0.3	5.0

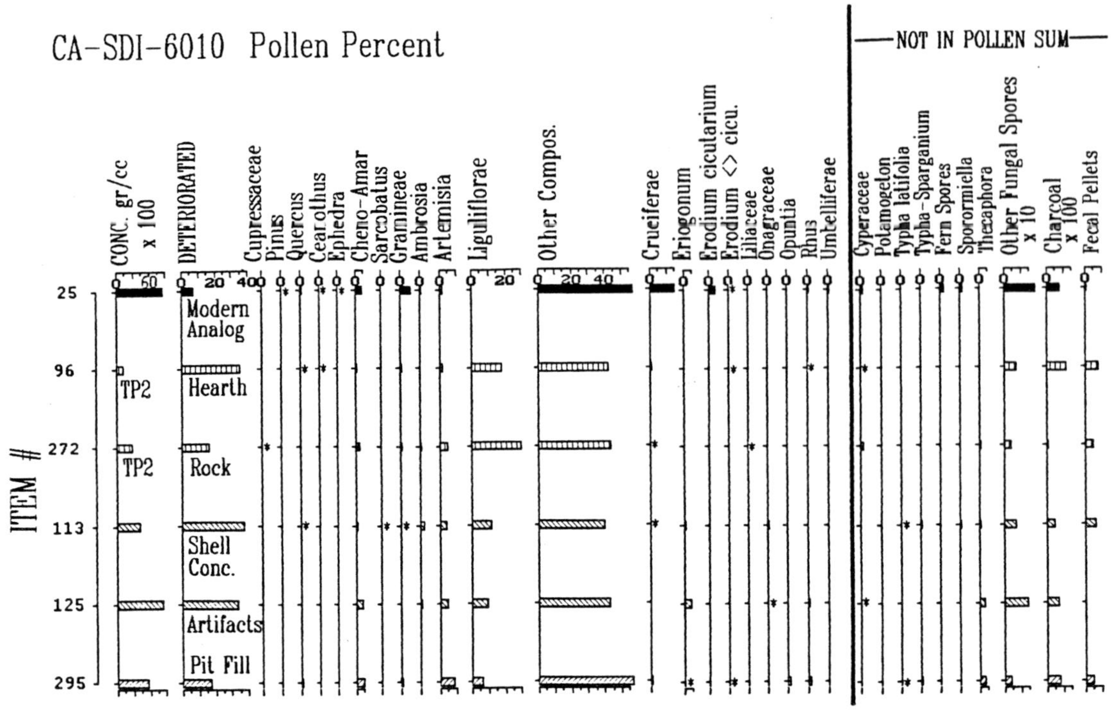

Figure 36. Percentage Pollen Diagram for CA-SDI-6010.
(* < =0.5%)

Items # 96 and # 272

These two samples (from Feature 1) contain the highest percentages of Liguliflorae pollen of the samples from CA-SDI-6010. The low concentration and high deteriorated percentages for Item # 96 probably indicate rapid infilling of the hearth with loose dirt in the site. Charcoal percentages (1190%) are the highest of any sample. The sedge (*Cyperaceae*) values (1.3%) in Item # 272 are nearly as high as those in the modern analog.

Items # 133 and # 125

Pollen concentration and percentages in these two samples are extremely similar. Low (<1%) values of cattail (*Typha latifolia* and *Tyhpa-Sparganium*) are present in Item # 113, the sample from an area with a high shell density.

Item # 295

This sample is different from the other archaeological samples. It has the highest *Artemisia* and Other Compositae percentages and lowest Liguliflorae percentages. It also has the only *Opuntia* (cholla) pollen recovered from the site, and has small (<1%) values of cattail (*Typha latifolia* and *Tyhpa-Sparganium*).

Interpretation

The pollen percentages, dominated by Other Compositae (sunflower family) pollen are sufficiently similar to the modern analog sample to indicate relatively little environmental change. The greatest difference can be explained by a transition in the weed flora from dominance by Liguliflorae in the archaeological samples, to prominence of Chenopodiaceae-*Amaranthus*, Gramineae, and Cruciferae in the modern analog. The overall similarity of the archaeological samples to the modern analog provides no clear-cut indications of ethnographic utilization of the features.

Environment at the Time of Occupation

Dominance by Other Compositae (sunflower family) pollen is a consistent feature of the pollen flora of low-elevation California west of the Sierra Nevada. At San Joaquin Marsh the high Compositae values are interrupted by Chenopodiaceae-*Amaranthus* dominance when salt marsh dominates the site (Davis 1989), but this is a local phenomenon. Dominance by Compositae pollen has persisted in the Central Valley since ca. 8,000 years ago (Davis 1990). The CA-SDI-6010 samples could date to anytime during the mid- to late-Holocene.

Item # 295 may date to a different period than the other samples. It contains fewer disturbance indicators and contains the only *Opuntia* pollen. Although the *Opuntia* could easily indicate ethnographic utilization, this type occurs in modern analog samples in low frequencies, so it could reflect more arid conditions than at present.

The pollen of aquatic plants is present in all of the archaeological samples. No pollen of willow and cottonwood occurs in the modern analog and archaeological samples, however. Interestingly, macrobotanical materials from SDI-6010 include high percentages of these two riparian species (see below). The sedge and cattail pollen probably does not come primarily from the San Luis Rey River vegetation. Their presence in the archaeological samples appears to indicate that the Whelan Lake basin supported aquatic vegetation.

Ethnobotanical Information

Although the pollen types recovered include plants that are used by humans, the percentages generally are too similar to background environmental levels to prove human activity. For comparison, Chenopodiaceae-*Amaranthus* at the center of CA-LAN-60 exceed 55 percent, while samples from the edge of the site have <10 percent Chenopodiaceae-*Amaranthus*. Elevated percentages of *Eriogonum* (4%) in Item # 125, the artifact concentration, could result from medicinal use of *E. fasciculatum* (Quillen et al. 1984), but all six samples contain some *Eriogonum* pollen, and such low percentages can be duplicated in the modern pollen rain. Likewise the *Opuntia* (1.3%) in Item # 295 (from hearth fill) could indicate cholla bud processing, but it could also be given an environmental interpretation.

Conclusions

Vegetation at the time of occupation of CA-SDI-6010 was little different from that of today, although the weed flora was dominated by Liguliflorae rather than introduced species. The presence of small percentages of sedge and cattail pollen in all samples indicates that the Whelan Lake basin was a marsh or lake. The pollen provides uncertain information regarding the diet of the occupants of CA-SDI-6010. The high Liguliflorae percentages and elevated *Eriogonum* and *Opuntia* values in Items # 125 and # 295 could reflect human utilization. But given the widespread occurrence of high Liguliflorae percentages in most California archaeological sites, Liguliflorae probably is a pre-settlement disturbance indicator. The percentages of *Eriogonum* and *Opuntia* are no higher than in the modern pollen rain, so they may represent minor environmental change rather than human utilization.

MACROBOTANICAL ANALYSIS

Charles H. Miksicek

Introduction

Five flotation samples from the lower levels of four test pits from the Whalen Lake site were analyzed. One sample was collected from Feature 1, a hearth which contained fire-affected rock, some charcoal, and shell. Two samples came from Feature 2, another hearth, containing small quantities of charcoal, shell, bone, lithics, and fire-affected rock. The remaining two samples represent general midden fill.

Several research questions related to the more general topics of subsistence and paleoenvironment may be addressed with charred botanical remains identified from these samples. These include:

(1) What plant resources were used by the occupants of CA-SDI-6010? Were these resources predominately upland, riverine, freshwater lacustrine, or estuarine? How large a catchment area is suggested by these plant resources? Was the site occupied seasonally and if so, what seasons?

(2) What was the local vegetation like during the occupation of the site? Was the site located above a river, lake, or saltwater lagoon? Was Whelan Lake a natural pond prior to the construction of an earthen dam?

Methods

All flotation samples were processed by Statistical Research personnel before they were submitted to this author for sorting and identification. During analysis, each sample was first passed through a nested series of geological sieves with mesh sizes of 4.0 mm, 2.0 mm, and 0.5 mm. This pre-sorting removes many modern rootlets and produces subsamples with similar-sized particles that are more effectively scanned by the analyst. Each fraction was then sorted at 7X under a binocular dissecting microscope into its component parts. Uncharred seeds were counted and identified to evaluate the modern seed rain and to provide a measure for the degree of bioturbation and mixing within samples. To provide an additional monitor on the amount of faunalturbation, rodent fecal pellets and insect parts were quantified on a relative scale ranging from 0 (none) to 5 (abundant). Charred seeds and other non-woody plant parts were identified and tabulated.

All pieces of charcoal large enough to manipulate were identified. Each fragment was first fractured to give fresh transverse and radial sections and then identified at 35X under a binocular dissecting microscope.

All seed and wood identifications are based on modern comparative material collected from southern California and standard seed and wood keys such as Musil (1963), Martin and Barkley (1973), and Miles (1978). Taxonomic nomenclature follows Munz (1963) and McMinn (1974). All botanical materials identified in the CA-SDI-6010 samples are listed in Table 41 along with ethnographic uses abstracted from Mead (1972) and Ebeling (1986). Seasonality data are derived from Munz (1963).

Raw data are reported in Table 42. This table includes sample volume before flotation, richness (the average number of charred seed taxa per liter), density (the average number of charred seeds per liter), the total weight of charcoal fragments larger than 2.0 mm, the bug part index, the rat pellet index, tallies of uncharred seeds, counts of carbonized plant remains, and identified wood charcoal totals.

Results and Discussion

Uncarbonized seeds are present in all analyzed flotation samples from CA-SDI-6010 in densities ranging from 0.5/liter to 23.2/liter (Table 42). It is most likely that all of these are recent intrusives and are not associated with the prehistoric use of the site. Plant materials are generally only preserved in open, archaeological sites if they are charred during cooking, storage, or use. This burning may be accidental or purposeful.

Charred seed densities range from 1.25/liter in Stratum II of Test Pit 10 to 17.25/liter near the bottom of Feature 2 (see Table 42). Grass seeds dominate the archaeobotanical assemblage and account for 34.7 percent of the non-woody plant remains. Four ethnographically important taxa (Mead 1972, Ebeling 1986) are present: native barley, maygrass, ryegrass, and sacaton grass. Small-seeded grasses such as three-awn, bluegrass, bentgrass, and lovegrass may represent minor foods or material accidentally included when other species were gathered using beater-baskets, breaking off seed heads, or by uprooting whole plants. Bromegrass and fescue grass occur consistently in California archaeological sites (Miksicek 1991) and probably represent important, but previously unrecognized, traditional food plants. All of these taxa would have been common in the grasslands of southern California prior to extensive historic disturbance. Depending on the exact species represented, these grass taxa could have been collected from early spring through mid-summer.

Spanish clover, true clover, and lupine seeds are present in three out of five samples from CA-SDI-6010. These plants may be used as edible greens and are available in the spring (Mead 1972, Ebeling 1986). Other key resources are goosefoot, chia, sugarberry, elderberry, and yucca. The monocot rhizome fragments noted in Table 42 may be either cattail or sedge, but it is difficult to identify charred roots without scanning electron microscopy. Most of the food remains from CA-SDI-6010 could be gathered from early spring through mid-summer. Sugarberries and elderberries are the only indicators for late summer residence and fall remains (nuts) are absent (although they may be found in unanalyzed samples from elsewhere in the site)

Table 41. Plant Taxa Identified from CA-SDI-6010.

Common Name	Scientific Name	Part Identified	Use	Season Indicated
Rose Family	Rosaceae	u	?	
Knotweed	*Polygonum* sp.	u	S	
Pickleweed	*Salicornia* sp.	u	G?	
Purslane	*Portulaca* sp.	u	G	
Pigweed	*Amaranthus* sp.	u	S,G	
Sedge	*Scirpus* sp.	u	S,R	
Morning Glory	Ipomoea or Convolvulusu	u	?	
Monocot Rhizome	Scirpus or Typha	w	R	early spring
Little Barley	*Hordeum pusillum*	s	S	spring
Fescue Grass	Festuca or Vulpia	s	S?	spring
Bluegrass	*Poa* sp.	s	S?	spring
Lupine	*Lupinus* sp.	s	G	spring
Spanish Clover cf.	*Lotus purshianuss*	s	G	spring
Clover	*Trifolium* sp.	s	G	spring
Chia	*Salvia* sp.	s	S	spring
Maygrass	*Phalaris* sp.	s	S	late spring
Yucca	*Yucca* sp.	w	G	late spring
Bentgrass-Type cf.	*Agrostis* sp.	s	S	early summer
Bromegrass	*Bromus* sp.	s	S	early summer
Ryegrass	*Elymus* sp.	s	S	early summer
Three-awn Grass	*Aristida* sp.	s	S?	early summer
Lovegrass	*Eragrostis* sp.	s	S?	early summer
Bedstraw	*Gallium* sp.	s	?	early summer
Sacaton Grass	*Sporobolus* sp.	s	S	summer
Mallow Family	Malvaceae	u,s	?	summer
Goosefoot	*Chenopodium* sp.	u,s	S,G	summer
Tarweed	Madia or Hemizonia	s	S	summer
Elderberry	*Sambucus mexicana*	u,s	F	late summer
Sugarberry	*Rhus* spp.	s,w	F,W	late summer
Cottonwood	*Populus fremontii*	w	W	
Willow	*Salix* spp.	w	M,W	
Cedar Family	Cupressus or Juniperus	w	W	
Saltbush Type	Atriplex or Suaeda	w	S,G,W	
Indian Buckwheat	*Eriogonum fasiculatum*	w	S,M,W	
White/Black Sage	*Salvia* spp.	w	S,M,W	
Composite Family	Asteraceae	w	?	
Yerba Santa	*Eriodictyon* sp.	w	M,W	

Key: u - uncarbonized seed, s - charred seed, w - charcoal, W - wood, ? - uncertain
S - edible seeds, G - edible greens, R - edible root, F - edible fruit, M - Medicine

Table 42. Plant Remains Recovered by Flotation from CA-SDI-6010.

Item #	270	127	147	289	291
Test Pit #	2	7	10	12	12
Stratum-Level	II-4	III-4	II-2	II-1/2	II-2.2
Depth (cm)	50-60	45-55	25-35	40-50	60-70
Feature #	1			2	2
Volume (liters)	4	4	4	4	4
Rat Pellet Index	0	0	0	1	0
Bug Part Index	1	1	1	1	1
Charcoal (gms.)	0.2	0.1	0.1	0.3	0.4
Uncarbonized (Intrusive) Seeds:					
Knotweed	1				
Pickleweed	1	1			
Elderberry	1		1		
Mallow Family	4		5	3	
Purslane	5		52	13	
Goosefoot	12		29	29	
Pigweed	9	1	87	29	1
Rose Family		1	8	17	1
Sedge			2		
Morning Glory		2		2	
Wood Charcoal:					
Cottonwood	1	3	1	11	15
Willow	1	2	2	8	11
Cedar Family	1			2	1
Saltbush Type	1				
Indian Buckwheat	2			2	
White/Black Sage	2			4	
Composite Family		1		12	13
Sugarberry		3		1	1
Yerba Santa				1	
Indeterminate	5	3	4	7	14

Table 42. Plant Remains Recovered by Flotation from CA-SDI-6010 (cont.).

Item #	270	127	147	289	291
Test Pit #	2	7	10	12	12
Stratum-Level	II-4	III-4	II-2	II-1/2	II-2.2
Depth (cm)	50-60	45-55	25-35	40-50	60-70
Feature #	1			2	2
Volume (liters)	4	4	4	4	4
Charred Seeds/liter	2.50	1.75	1.25	6.75	17.25
Charred Taxa/liter	1.50	1.00	1.00	3.75	3.50
Carbonized Seeds (etc.):					
Sacaton Grass	1				
Three-awn Grass	1				1
Lovegrass		2			
Little Barley				4	15
Maygrass				1	4
Fescue Grass				2	1
Bluegrass				1	1
Bentgrass-Type				1	
Bromegrass					1
Ryegrass					5
Goosefoot		1		1	1
Lupine			1	2	7
Spanish Clover				2	4
Clover				1	6
Tarweed					2
Chia				1	
Bedstraw				1	
Mallow Family				1	
cf. Sugarberry	2				
Elderberry					1
cf. Yucca Stem	2	2	1		
Monocot Rhizome			2	6	
Seed Fragments	1			1	20
Vesicular Material	3	2	1	2	
Charred Miscellaneous Remains:					
Grass Stems				1	1
Termite Pellets		1	7	1	2

How do you distinguish between remains charred by people during processing activities from those burned by natural grassland fires? The residue from natural fires usually includes a greater diversity of plant parts including stems, floral parts, leaves, and other vegetative material. Wildfires also leave behind a greater range of species many of which have no ethnographically documented uses. The archaeobotanical assemblage from CA-SDI-6010 is dominated by charred seeds from ethnobotanically important food taxa, so it is most likely that these remains are cultural and not natural.

The charred seed assemblage from CA-SDI-6010 suggests that a mosaic of native grassland and coastal sage scrub existed around the site during the time it was occupied. Wood charcoal provides additional clues for paleoenvironmental reconstruction. Over half of the identified charcoal (53.9% of 102 fragments, see Table 42) comes from riparian trees such as willow and cottonwood. Cottonwoods and willows were probably common along the floodplain of the San Luis Rey River in the past. An additional 41.2 percent reflects coastal scrub or chaparral taxa such as Indian buckwheat, sage, composites (sunflower family), sugarberry, or yerba santa. These shrubs probably would have been common on the slopes above and to the west of CA-SDI-6010.

Based on this mixture of seeds and charcoal, it seems unlikely that CA-SDI-6010 was located near a brackish lagoon at the time it was occupied. A single fragment of saltbush-type charcoal is the only example of a halophyte that would have been found near a salt marsh. What would a charred plant assemblage look like from a site situated near a brackish lagoon? Plant remains from other sites near the southern California coast provide a few clues. Table 43 summarizes archaeobotanical data from seven other sites in this region including CA-ORA-225 and 389 (Miksicek 1990), CA-LAN-47 (the Admiralty Site, Brooks and Johnson 1991), CA-LAN-60 (Miksicek, unpublished data), CA-LAN-229 (Talepop, Hammett 1990), CA-SBA-1816 (Hammett 1989a), and CA-SLO-165 (Hammett 1989b). Sedge seeds, indicative of fresh to brackish water marshes are present in four sites in Table 43. Annual saltbush, a plant that thrives in wet to dry, alkaline or saline soils is also listed from three sites. Pickleweed, a good saltmarsh indicator, is only reported from CA-LAN-60. Cordgrass is present at CA-ORA-389 and CA-SLO-165. Cordgrass, a salt marsh species, is found in nearly pure stands along the southern California coast (Beauchamp 1986:70). It produces an edible grain very similar to wild rice (*Zizania aquatica*) and would have been an important food in the past. The recovery of significant numbers of the aforementioned taxa from CA-SDI-6010 would be necessary to suggest that it was located near a brackish water lagoon at the time it was occupied.

There are some indications that a freshwater marsh or lagoon was located near SDI-6010. The moisture-loving taxa actually identified from the site are cottonwood, willow, and cattail/sedge. Cottonwood and willow are riparian species that do not survive in waterlogged soil. They are most commonly found along perennial or semi-permanent streams, or occasionally on raised hummocks (with better drained soils) in marshy areas. They do not tolerate saline soils. The two species would have been common along the San Luis Rey River or its tributary drainages. The most mesic plants in the entire collection are the cattail or sedge rhizome fragments. These plants could have been found in a variety of habitats ranging from seasonal pools and perennial streams to lagoon margins and marshland. They thrive in waterlogged soil, such as found in a marshland environment, however. Like willows, cattails do not like saline soils.

Table 43. Charred Plant Remains from Sites Along the Southern California Coast

County	SDI	ORA	ORA	LAN	LAN	LAN	SBA	SLO
Site Number	6010	225	389	47	60	229	1816	165
Number of Samples	5	15	11	67	11	8	113	7
Volume Range (liters)	4	1-2	1	NA	2-5	50-100	6-25	50
Charred Taxa/liter	2.15	2.96	2.91	NA	2.05	0.57	6.09	0.20
Charred Seeds/liter	5.90	4.35	7.03	NA	3.87	22.16	3.26	4.55
Bromegrass	+							
Lovegrass	+							
Sacaton Grass	+							
Three-awn Grass	+				+			
Bluegrass	+	+			+			
Fescue Grass	+	+			+			
Little Barley	+	+	+		+	+	+	+
Maygrass	+	+	+		+	+	+	+
Ryegrass	+	+	+					+
Bentgrass	+	+			+			
Hairgrass		+			+			
Cordgrass			+					+
Panic Grass				+	+			
Lupine	+				+			
Clover	+		+		+			
Spanish Clover	+	+	+		+		+	
Bedstraw	+		+		+	+	+	
Chia	+				+	+	+	+
Goosefoot	+	+			+	+	+	+
Tarweed	+							+
Mallow	+	+		+				+
Borage		+			+	+	+	
Red Maids		+				+	+	
Knotweed			+		+	+	+	
Pigweed			+	+	+	+		
Dock				+		+	+	+
Sedge				+	+	+	+	
Sunflower				+		+		
Peppergrass					+	+		+
Annual Saltbush					+	+	+	
Pickleweed					+			
Coyote Tobacco							+	
Elderberry	+					+		
Sugarberry	+	+				+	+	
Toyon		+				+		
Walnut		+	+			+		
Juniper				+	+	+		
Acorn Shell						+	+	+
Islay						+		+
Manzanita						+	+	
Monocot Rhizome	+				+			
Monocot Bulb						+	+	
Yucca	+					+		+

Two plant taxa identified from CA-SDI-6010, "cedar family" charcoal and yucca, seem out of place in the Oceanside area. Both are more characteristic of the interior (desert) side of the southern coast ranges. Although juniper and cypress are difficult to distinguish by wood anatomy, the "cedar family" charcoal is probably not from cypress, as the only two known species from the area (Cuyamaca cypress [*Cupressus stephensonii*] and Tecate cypress [*C. forbesii*]) are found in south-central San Diego County near the Mexican border (Beauchamp 1986:40). California juniper (*Juniperus californica*) is an occasional element along drainages on the west flank of the San Gabriel mountains of Los Angeles County (Hanes 1976:69). Both cypress and juniper (as well as coast redwood) are part of the Rancho La Brean flora dated between 40,000 and 25,000 years B.P. (Warter 1976). It is quite possible that several coniferous species had slightly more extensive ranges in the more recent past. Carbonized juniper seeds have been identified from three sites listed in Table 43. Pine charcoal has been identified from CA-ORA-225, miles from any known modern stands of the trees (Miksicek 1990). The "cedar family" charcoal from CA-SDI-6010 may not necessarily represent material imported from long distances or driftwood carried by the San Luis Rey River. The yucca stem fragments may also reflect a slight extension of the known modern range of the genus. Today *Yucca whipplei* gets to the Pacific coast at Encinitas, only 18 kilometers (11 miles) south of the site (Beauchamp 1986:43). Until we know more about the prehistoric distribution of these taxa, it is impossible to decide whether the CA-SDI-6010 macrofossils represent distribution changes or the long distance importation of economically important plant resources.

One more point should be mentioned regarding the CA-SDI-6010 archaeobotanical assemblage. A similar mixture of grasses, legumes, and small edible seeds has been identified from many sites in southern California (see Table 43) and elsewhere throughout the state (Miksicek 1991). This "Grass-Legume-Small Seed Complex" has strong parallels with the types of remains found in late Epipaleolithic (or early Neolithic) sites in the Near East (Miksicek 1991). It even shares some species or genera with the "Eastern Agricultural Complex" found in Late Archaic and Early Woodland sites in eastern North America (Miksicek 1991). For example, native barley has been considered an incipient cultivar in both eastern North America (Asch and Asch 1985) and southern Arizona (Adams 1987). Maygrass has also been examined as a native plant in the early stages of domestication in the Southeast (Asch and Asch 1987). Perhaps as we gather more archaeobotanical data from California sites, and look carefully at taxa such as native barley, maygrass, ryegrass, cordgrass, goosefoot, chia, Spanish clover, coyote tobacco, and tarweed in terms of both size and morphology, we may come to realize that native Californians were more than simple collectors of "wild" plants.

Summary

1. The archaeobotanical assemblage from CA-SDI-6010 contains a number of ethnographically important taxa including native barley, maygrass, ryegrass, Spanish clover, lupine, chia, goosefoot, tarweed, elderberry, sugarberry, and yucca. This complex is similar to that found in many other sites along the southern California coast.

2. These macrofossils suggest that the site was occupied predominately from early spring through late summer.

3. Most of these plant resources could have been gathered in the grassland or coast scrub communities found around the site prior to recent historic disturbance.

4. The wood charcoal assemblage indicates a riparian woodland along the San Luis Rey River with willows and cottonwoods. The coastal sage scrub community is also reflected in the charcoal record. There is almost no archaeobotanical evidence for a brackish water lagoon in the immediate vicinity of

the site at the time of occupation. The occurrence of cattail and sedge in the assemblage suggests that a freshwater marsh or lagoon may have been located nearby.

5. Yucca stem remains and "cedar family" charcoal recovered from the site may indicate a more extensive range for these two taxa in the past or the long distance importation of economically important taxa.

CONCLUSIONS

In most respects, the results of micro- and macrofossil analyses for SDI-6010 agree on the type of environment present in the Whelan Lake area during the occupation of SDI-6010. Neither analyses found evidence for the occurrence of a nearby brackish lagoon. Given the presence of cattail and sedge in both micro- and macrofossil assemblages, it is likely that a freshwater marsh or lagoon occurred close to the site.

The absence of pollens of willow and cottonwood is puzzling, as a high percentage of wood charcoal from SDI-6010 is from these two species. Maybe willow and cottonwood did not occur along the San Luis Rey River near the site. In that case, firewood from these species would have been collected from further away or was found near the site in the form of driftwood.

The archaeobotanical assemblage of SDI-6010 is dominated by a similar mixture of edible grasses, legumes, and small seeds found at many other sites in southern California. Possibly, rather than being "wild" plants, some of these species may have been deliberately encouraged to grow in disturbed areas, thus representing incipient cultivars. The presence of an associated disturbed area during the site's occupation is indicated by high percentages of Liguliflorae pollen.

Most of the plant food remains from the Whelan Lake site could be gathered from early spring through mid-summer, whereas fall remains (e.g., nuts) are absent. Of course, rather than suggesting that the site was only occupied during a certain time of the year, this only specifies when plants formed a major portion of the diet of the site's inhabitants.

CHAPTER 11

ANCIENT SETTLEMENT AND SUBSISTENCE ALONG THE SAN LUIS REY RIVER

This chapter represents the integration of the various pieces of information obtained from field work at the Whelan Lake site (CA-SDI-6010). The chapter is organized from specific to general. First, site specific characteristics are reviewed. Site composition is evaluated by examining layout, stratigraphy, feature types, and cultural deposits. Results from the various analyses of collections from the site are correlated. Distributional covariation of various artifact and ecofact classes is appraised. The subsistence base of the site is evaluated. Within a temporal context provided by radiocarbon dates, the results of the paleobotanical analysis is integrated with geomorphological evidence to characterize the local paleoenvironment.

Once the discussion of site specific characteristics is completed, regional trends are identified. The San Luis Rey River basin is divided into its various environmental components, with each physiographic province described. Representative sites from each physiographic zone are compared, both within and between the individual zones. Questions asked are: (1) was occupation within a zone homogeneous or did it vary between sites in specific ways and for specific reasons, and (2) did settlement between zones differ, and if so, in what manner?

After regional trajectories have been identified, data from throughout the basin is used to construct a model of settlement along the San Luis Rey River. Although the time period represented at the Whelan Lake site is emphasized, later events are also examined. The model is tied to environmental changes occurring during the archaeological sequence, and assesses how these changes affected settlement in the basin. Potential cultural factors at play are also evaluated. The synthesis concludes with a critical discussion of data gaps in the model and directions for future research.

THE WHELAN LAKE SITE (CA-SDI-6010)

The Whelan Lake site is located on a low knoll overlooking the San Luis Rey River 200 m to the southeast. The mouth of the river is 7 km southwest of the site. The knoll forms part of a late Pleistocene river terrace and its elevation of 20 m AMSL kept the site safe from even the most substantial flood episodes. Although the site has been impacted by irrigation activities, archaeological integrity is relatively good.

Temporal Context

Six radiocarbon assays of marine shell valves (*Chione* sp.) were conducted. The samples originated from various horizontal and vertical proveniences in the midden. The resulting ^{14}C dates are well-clustered, ranging between 7,400 and 6,500 B.P. With ^{13}C and reservoir effect adjustments applied in tandem, a near cancellation of the adjustments occurs. Therefore, in our discussions pertaining to radiocarbon determinations for SDI-6010, we use uncorrected, conventional ^{14}C readings.

Five obsidian flakes were submitted for trace-element analysis to determine their source, and subsequently examined to determine hydration rind thickness. Three of the samples originate from the

Coso source, and for these absolute dates were extrapolated from Meighan's (1978) Coso obsidian index of 220 years/micron. The resulting dates, 354 B.C., A.D. 196, and A.D. 494, although mutually fairly consistent, do not match the radiocarbon dates. Various factors may have been responsible for small, incompatible hydration measurements. We believe that the obsidian samples lack integrity, and thus the hydration dates are ignored in our evaluation of the site.

If the radiocarbon dates are correct, the site was occupied during the La Jolla I period. The occurrence of a Pinto style projectile point in the midden fill is consistent with an early La Jollan occupation. The occurrence of a few Tizon Brown sherds on or near the midden surface implies some minor use of the site during the San Luis Rey II period.

Site Layout

SDI-6010 is a roughly oval-shaped shell midden, measuring 90 m by 60 m. The midden fill is 1 m thick at the center but thins rapidly towards the margins of the site. The central portion of the midden contains an area of intensified shell density that has a similar oval form, measuring 56 m by 38 m. Within this area of higher shell concentration, four archaeological features were identified. In addition, four clayey areas were discerned that represent cultural activities, albeit of an unknown nature.

Feature Types

All features found at the Whelan Lake site are interpreted as hearths. Two types of hearths were defined. The first type, represented by Feature 1, is a circular pit filled with fire-cracked rock, compacted ashes, charcoal, shell (mostly *Donax*), and burned animal bone. Two manos were located near the edge of the feature. A radiocarbon date obtained from *Chione* occurring near the bottom of the feature suggests that the hearth was used approximately 7,060 ± 100 B.P.

Features 2, 3, and 4 represent a second type of hearth. Presumably circular pits, these features are characterized by only a few fire-cracked rocks, small amounts of ash, but abundant charcoal. Only Feature 2 of this type was excavated. On the edge of this pit a small cluster of fire-cracked rock occurred. Similarly to Feature 1, the fill contained an increased density of *Donax* compared to midden fill above it. A *Chione* sample from the feature fill provided a radiocarbon date of 6,540 ± 90 B.P.

Without further excavation it cannot be determined whether the two types represent functional differences. It may be that Feature 1 was fairly undisturbed, with both hearth and adjacent activity area sealed by subsequent deposition events. Features 2, 3, and 4 with their sparse fire-cracked rock fill may represent abandoned hearths robbed of their fill.

Clayey Areas of Unknown Cultural Origin

Within the midden, four areas contain high clay content that is organically enriched. These pockets of atypical midden fill are characterized by high shell densities and small charcoal fragments. Features 2 and 4 occur in such areas. Consistently high ecofact content and associated hearths suggest that the clayey areas are cultural. At this point, interpretation is confined to juxtaposing a number of optional scenarios of clay introduction to the site. Clay may have been brought in adhering to imported shell, suspended in water for food processing, or intentionally for construction purposes (see Chapter 5). Further work at the site should consider these options.

Distributional Patterns

The test pits excavated at SDI-6010 provide data on intrasite distributional patterns. Table 44 summarizes artifact counts and ecofact NISP per cubic meter for each test pit, differentiating between chipped stone, ground stone, shellfish, fish, terrestrial fauna, and waterfowl. Frequencies have been adjusted to conform to a 1 m^3 unit, and do not reflect actual counts of materials present in the individual test pits.

Table 44. Artifact Counts and Ecofact NISP per Level for Each Test Pit.

Test Pit	Terrestrial Fauna	Waterfowl	Shellfish	Fish	Ground Stone	Chipped Stone
1	51.1		57.8	2.2		28.9
2	137.8	1.0	547.9		5.1	87.8
4	10.0		252.0		2.0	38.0
5	7.1		145.7			21.4
6	61.5	1.5	558.5			89.2
7	42.9	1.4	645.7			24.3
8	599.4	0.6	516.9	3.7	1.3	144.4
9	50.0		187.1			124.3
10	175.5		797.8	2.2		155.6
11	150.0		353.3			143.3
12	100.0		817.5		7.5	102.5
Avg.	125.9	0.4	443.6	0.7	1.4	87.2

Not surprisingly, the site center, exemplified by Test Pit 8, has the greatest diversity and density of cultural material. The high frequency of terrestrial animal bone in Test Pit 8 is dramatic, with an NISP count exceeding that of shell.

Shellfish is most abundant in Test pits 7, 10, and 12. Interestingly, these are the three test pits excavated in three of the four clayey areas, lending further credibility to the argument that cultural factors are responsible for importation of the clay.

Ground stone is fairly rare at the Whelan Lake site. It is most abundant in Features 1 and 2, the two hearths excavated with Test Pits 2 and 12. The frequency of bone is not directly correlated with the ground stone, which is somewhat surprising because the faunal material was so fragmented.

Chipped stone, present in moderate numbers in the hearths, is most profuse in the central and southwestern portion of the midden. Although high shell frequencies are often accompanied by an increase of chipped stone, the correlation is not consistent, and there is no evidence for a functional correspondence between the two classes. Such a correlation does exist between chipped stone and faunal remains, with ratios between the two classes consistently uniform for all test pits.

General Site Constituents

An appreciation of general site constituents is a prerequisite for intersite comparisons. Table 45 presents the percentage contribution of various ecofact classes to total NISP at SDI-6010. Of course, this table denotes the material preserved in the archaeological record, and is not an exact reflection of the economic importance of these various faunal categories. A rough idea of the relative value of meat obtained from discrete ecological zones, however, can be gleaned from these percentages.

Table 45. NISP Percentages of Ecofacts at the Whelan Lake Site.

Terrestrial Fauna	22.06 %
Waterfowl	0.07 %
Fish	0.12 %
Shellfish	77.74 %

Lithics form a large portion of cultural material recovered from SDI-6010. For ground stone, the average count per cubic meter is 1.4 items, contrasting with 87.2 fragments for chipped stone. As noted in Chapter 7, lithics collected from the site consist of 704 flaked stone items and 15 ground stone pieces. Most (96.28 %) of the chipped stone consists of small debitage, with the remainder composed of cores (0.43 %) and tools (3.28 %). For comparisons with other sites, the tool category is the most critical (Table 46). At the Whelan Lake site, the lithic assemblage was dominated by milling stones and heavy-use scraping, chopping, and hammering tools. Formal, bifacial tools constituted just over 10 percent of the collection.

Table 46. Lithic Tool Ratios for the Whelan Lake Site.

Milling stones	39.48 %
Scrapers, planers, and choppers	23.68 %
Hammers	26.32 %
Bifacial tools	10.52 %

Conclusion

Examination of features and cultural materials from the Whelan Lake site indicate that the occupants were sustained by a mixed economy of estuarine and terrestrial resources. In volume, shellfish comprises the bulk of faunal remains present in the midden. Yet, shellfish NISP is only 3.5 times larger than the NISP for small terrestrial mammals, highlighting the importance of non-aquatic sources for meat in the diet of the early La Jollan inhabitants of the site. The stone tool assemblage of SDI-6010 further typifies a coalescence of multiple adaptational strategies. Chipped stone debitage is most abundant in areas with increased densities of animal bone, a correspondence not found between ground stone and faunal remains. These correlations may indicate that lithic tool manufacture was associated with meat preparation, whereas milling stones were mostly used for plant processing. It also suggests that the shell midden contains identifiable activity areas, rather than being a homogeneous mass of debris. This interpretation is further supported by the two excavated hearths, each associated with a high occurrence of ground stone and an increased density of *Donax* shell species. Future work at the site needs to examine the extent of spatial variation, and attempt to identify activities specific to culturally discrete areas in the midden.

REGIONAL TRENDS IN THE SAN LUIS REY RIVER BASIN

Introduced and framed by a review of the environmental history of the San Luis Rey River, cultural events in this setting are linked to ecological shifts through time. A model of cultural evolution is posited that accounts for changes in settlement and subsistence observed in the archaeological record.

Physiography and Geography

The San Luis Rey River basin can be divided into lower, middle, and upper portions, each possessing distinct physiographic zones (Figure 37). The lower basin stretches from the Pacific Ocean to Bonsall. This portion of the drainage includes sandy beach at the open coast, estuary, freshwater marshland, and alluvial valley. The middle basin is defined as that portion of the drainage between Bonsall and the Lake Henshaw dam. It is comprised of a lush grassy valley, extensive oak groves on the foothills and mountain slopes, and pines and cedars on the top of Mount Palomar. The upper basin is that section between Lake Henshaw the the river's headwaters in Bucksnort Mountain. This section is dissected by various tributary channels cutting through inland valley, hills, and mountains. To provide an environmental context for modeling cultural evolution in the San Luis Rey River basin, each of these ecological zones is discussed. Where possible, the discussion will focus on environmental changes through the Holocene and how these changes altered the various subsistence sources available to the inhabitants of the basin.

The Lower Basin

The lower basin of the San Luis Rey River drainage covers a distance of 19 km. The gradient for this distance is less than 2.5 m/km. Currently, this portion of the basin is a wide, alluviated valley surrounded by low, gently sloping hills and ridges above the Late Pleistocene river terrace. Since the early Holocene, some dramatic ecological changes have occurred in the lower river valley.

Coastal Area

Compared to the coastal portions of northern and central California, The littoral of southern California is decidedly less rugged. Most of the land between mountain and shore consists of uplifted marine terraces and wide sediment-filled basins, forming a relatively flat, unbroken coastal plain 5 km to 60 km wide between southern Ventura County and the Mexican border. As in northern and central California, the south littoral includes stretches of rocky, sandy, and protected (estuarine and lagoon) coast. Because of its extensive lowlands, the south coast contains a large number of estuaries and lagoons. Widespread modern development and sea-level rise have made many of these inlets unrecognizable or, in many cases, entirely obliterated them.

Any study of temporal changes in coastal environment must consider the effect of sea-level rise. After the last glacial, during the Late Pleistocene and Early Holocene, accelerated sea level rise altered the California littoral significantly. In California, the amount of land submerged between the last low sea-level stand of about 120 m below the present position (ca. 17,000 B.P.) and stabilization (ca. 5,000

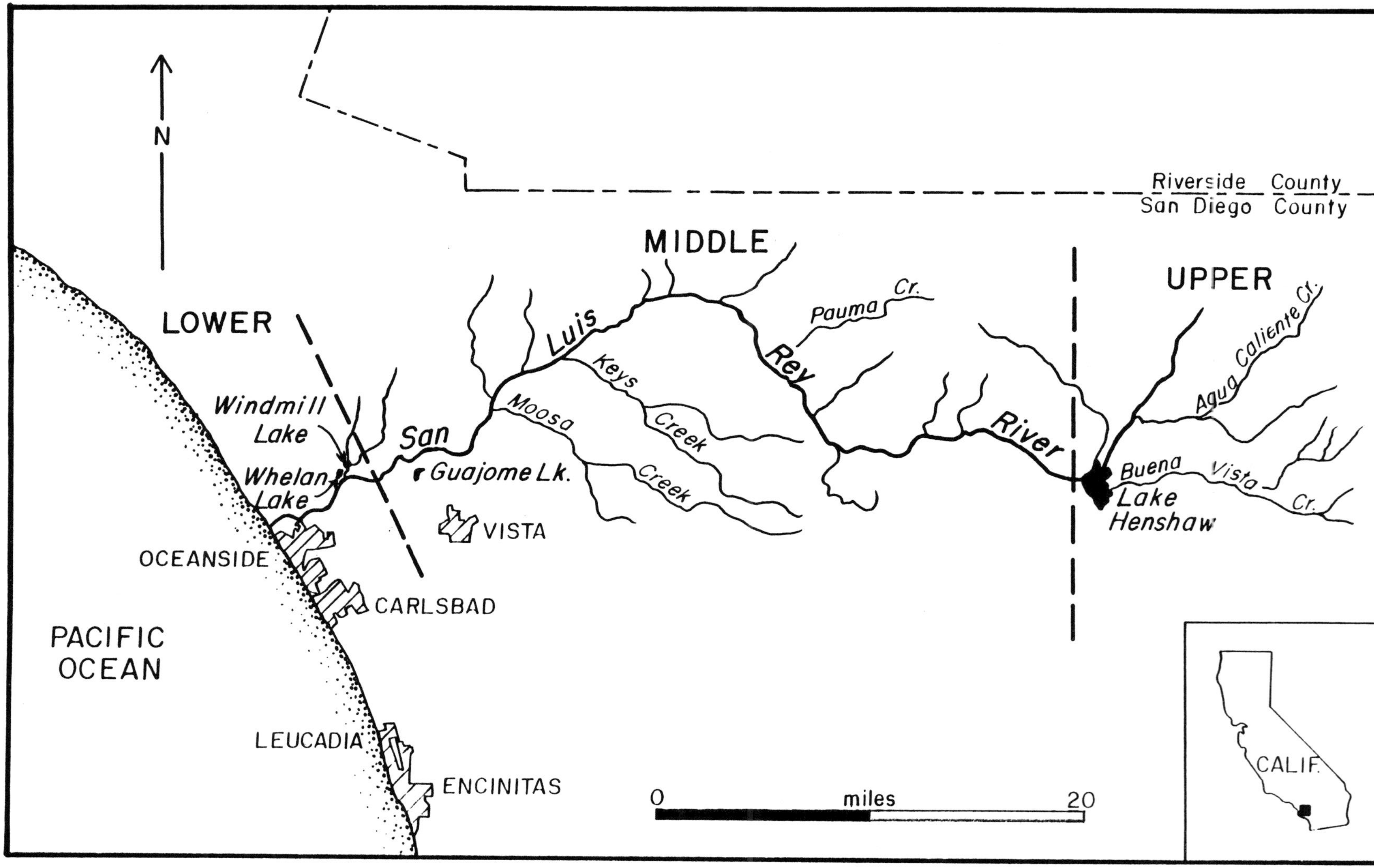

Figure 37. Map of the San Luis Rey River basin with major physiographic provinces.

B.P.), is indirectly correlated with latitude (Bickel 1978). Generally, the submerged coastal shelves are widest in the north (ca. 40 km) and become progressively narrower as latitude decreases. In San Diego County the width of the coastal shelf submerged during that time period ranges between 5 km and 17 km. Not until after the most recent rise of sea-level, the Flandrian transgression (ca. 5,000 B.P.), did stabilization (in the form of a more gradual rise of 10 cm per century) occur (Inman 1983).

At the mouth of the San Luis Rey River, a 9 km width of littoral shelf has been inundated since 17,000 B.P. (Jones 1991:Figure 3). Most of the inundation occurred between 12,000 B.P. and 5,000 B.P., when about 8 km of coastal land was submerged (Bloom 1983). Consequently, no direct physical evidence has survived of the Early Holocene open coast. No doubt, the habitat was a relatively harsh and unprotected environment, lacking fresh water and being inhospitable to permanent habitation.

The Estuarine Zone

The rise of sea level had a tremendous impact upon drainage patterns, which in turn affected the types of species inhabiting the littoral. The particular composition of the California shelf resulted in an efflorescence of estuary and lagoon development between 8,000 and 5,000 years ago (Carbone 1991; Emery 1967; Erlandson 1988). The rocky foreshores that supported limited shellfish fauna changed to estuarine and lagoon configurations which fostered an increase in marine, avian, and small terrestrial species. Around these estuaries, a variety of ecological zones were established (Desgrandchamp 1976; Yesner 1980), allowing new flora taxa to be introduced into the local plant communities.

An estuary is a transition zone where fresh water and sea water mix, producing a unique environment with physical conditions unlike those of neighboring riverine and oceanic systems (Kennish 1986). It receives a steady supply of nutrient materials washing down rivers and flushing out of bordering marshlands and is a rich and thriving environment. Under the right circumstances, this habitat will host a wealth of marine fauna. An ongoing process of reshifting of the ecological balance marks this milieu, however. The characteristics of the water in estuaries are in a constant state of flux due to tidal forces, river flow, and winds. The principal variable in most estuaries is salinity. Usually there is a continuous gradient between the fresh river water and the influence of the tides, as well as river run-off and weather conditions (Wolff 1983:151). Another variable influenced by the environment is the temperature of the water. Oxygen levels can vary from severe shortages to full saturation and during extreme droughts desiccation can destroy many of the inhabitants of the estuary. Finally, in addition to normal flushing cycles, during extreme flooding entire estuaries can be washed out, leaving few faunal resources for local human exploitation (for in depth discussions of these and other variables see Dyer 1973; Jones and Wolff 1980; Kennish 1986; Ketchum 1983; Knox 1986).

Organisms of the estuary have developed a wide variety of mechanisms to adapt to the changing conditions. For example, fluctuating levels of fresh and salt water entering the estuary cause considerable stress on the organisms that inhabit the environment. Some species migrate out of the area during stressful times, whereas others have adaptive mechanisms to deal with such stresses (e.g., bivalves can close tightly to survive periods of low salinity). Many, however, simply perish during these periods, being reintroduced later during favorable conditions.

The maximum extent of the estuarine zone during the Early Holocene florescence of lagoon development is unknown for the San Luis Rey River. Geomorphologic study of the area suggests that an estuary or brackish lagoon once extended up to 2 km inland (see Chapter 2). Two constraints appear to have prevented the formation of the kind of extensive inlet exemplified by, among others, the Batiquitos Lagoon and the San Diego River Lagoon to the south. The first constraint results from the particular composition of the stream channel just above the mouth of the river, from about 0.75 km to 2 km inland (see Figure 5). This lower stretch of the river (aptly named Oceanside Narrows), is a fjord-like strait, incised in Tertiary bedrock. This restriction is at its narrowest at a point 2 km from the mouth. Here, the present floodplain is not more than 180 m wide. Without the alluvium, which has accumulated to 3-5 m above sea level, the floodplain would be a little more than 150 m wide. In such a narrow constriction a salt water lagoon, even if it penetrated farther upstream, would become obstructed by tidal and fluvial sediment. The resulting lagoon would soon become an estuary. Only during major floods would the estuary, including its sediments and fauna, be flushed seawards.

The second constraint has to do with the elevation of the river valley hosting the ancient lagoon. Upstream from the Oceanside Narrows, the floodplain widens into a broad basin. Within this basin, the river gradient is low, approximately 2 m/km. The outer extent of the river flood plain is best defined by the 60 ft contour line flanking the river as depicted on the 7.5 minute topographic map for the area, the Oceanside Quadrangle (see Figure 6). The region confined within this contour line corresponds largely to the area of Holocene alluvium. The deep, fluvial deposits in the basin accumulated during successive floods of late prehistoric and historic times. Prior to that, extensive and thick alluvium was deposited in the area around the Mission San Luis Rey during the Flandrian Transgression, which culminated circa 5,000 B.P. (see Chapter 2). The depths of these deposits, and thus the depth of the river basin at the time the lagoon was established (8,000 to 5,000 B.P.), is unknown. From 7,400 to 6,500 B.P. (the time of occupation for the Whelan Lake site), local sea level was between 10 m and 15 m below the current position (Figure 38). Consequently, deposits 5 km inland (the 30 ft contour line) must have been at least 25 m thick if an ancient lagoon once was located there. This is highly unlikely. According to Antony Orme (personal communication 1991), the furthest inland extent of sea water may have been about 4 km. This maximum extent, if it ever occurred, would have been short-lived, however, soon becoming brackish and eventually (within 200 years) fresh water.

With adequate data lacking on the lagoonal history of the San Luis Rey River, it is useful to examine Batiquitos Lagoon, located about 15 km to the south. Based on radiocarbon dated shell from core samples obtained from adjacent archaeological sites, Miller (1966) has chronicled the various episodes of opening and closing of Batiquitos lagoon. In this reconstruction, the condition of the lagoon, deep water versus mud flat/shallow estuary, is based on the weight ratio of *Chione* and *Argopecten* in the samples. *Argopecten* need to be covered by water at low tide and prefer an open lagoon with good tidal flushing. *Chione* are more sturdy and survive in mud flats. During the earliest occupation of Batiquitos Lagoon, over 8,000 B.P., *Argopecten* was the dominant species followed by *Chione*. By 7,000 B.P., *Chione* began to dominate *Argopecten*, thereby suggesting a lapse of deep, open

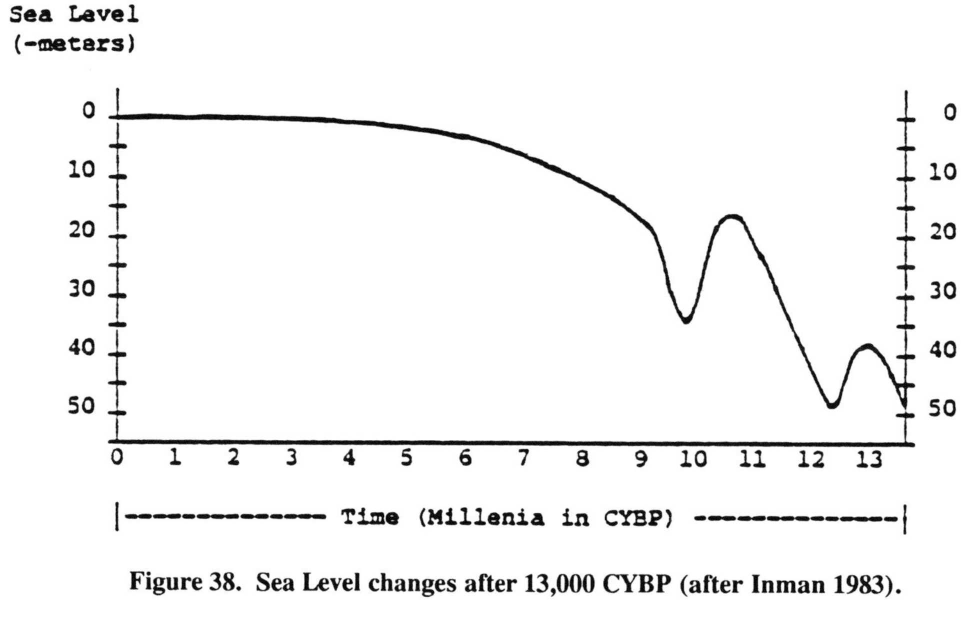

Figure 38. Sea Level changes after 13,000 CYBP (after Inman 1983).

lagoon conditions. This hypothesis is consistent with the slowing of sea level rise at this time. By 5,200 B.P., *Chione* was dominant, and remained so until the sites were abandoned circa 3,500 B.P. The abandonment of the area is correlated with the silting of the lagoon circa 3,500 B.P. A scenario can be reconstructed, showing the lagoon open from 10,000 to 3,500 B.P., closed from 3,500 to 1,500 B.P., reopened from 1,500 to 800 B.P., and closed again from 800 B.P. until present.

The physiographic character of the coastal portion of the lower San Luis Rey Basin is markedly different from that of Batiquitos Lagoon and most other estuaries on the California coast. Instead of a broad body of water forming the wide mouth of a minor river (such as at Batiquitos Lagoon), the ancient San Luis Rey River lagoon had a narrow inlet, much more susceptible to sedimentation. Even though events pertaining to changing conditions in Batiquitos Lagoon can not be simply transposed to the lower San Luis Rey River in the Early Holocene, some comparison is possible. The time of initial lagoon formation and apex of deep(er) water conditions were probably the same for both estuaries. So far, no sites earlier than 7,800 B.P. have been recorded for the San Luis Rey River drainage, but it is clear that after 8,000 B.P. the mouths of both the Batiquitos and San Luis Rey rivers were occupied by littoral food collectors. Although *Argopecten-Chione* ratios for the Batiquitos Lagoon can not be directly compared to those at sites in the lower San Luis Rey River basin, they may provide a rough idea of the degree of sedimentation in the San Luis Rey River lagoon. The Whelan Lake site (SDI-6010), dated from 7,400 to 6,500 B.P., has a shellfish assemblage composed of 83 percent *Chione* and six percent *Argopecten*. This ratio is roughly similar to that of SDI-5030, occupied between 6,300 and 4,000 B.P., and located a few km downstream from SDI-6010. In none of the other recorded shell assemblages in this basin does *Argopecten* dominate *Chione*, suggesting that throughout its existence the San Luis Rey River lagoon was shallow and characterized by sand or mud flats.

In sum, various lines of evidence indicate that an estuarine zone occurred in the lower San Luis Rey River basin. This estuary would have been most extensive between 8,000 and 5,000 B.P. After 5,000 B.P. silting and alluviation set in, eventually resulting in closure. Not until we possess data obtained from core samples in the alluviated lower basin will we be able to determine the extent of the ancient lagoon. Such a sampling program should be accompanied by radiocarbon dating of the samples.

Freshwater Marshland

Besides promoting lagoon development, major rises in sea level also occasioned inland alluviation. Thick alluvial deposits in the area around the San Luis Rey Mission, including the region near Whelan Lake, raised ground level significantly, thereby decreasing the gradient of the river below Pala. When the estuary was in existence, freshwater marshes would have formed further upstream as waters backed up behind the lagoon. Such was the case not only in the San Luis Rey drainage, but in most drainages in southern California (True 1966:59). Whelan Lake was marsh-like before damming in 1946 and it is likely that these conditions extend well into prehistory (see Schmidt et al. 1992). Apparently, the Whelan Lake site was situated in a favorable location: on a knoll overlooking the river, safe from river floodings, in an area where extensive marshland and surrounding foothills guaranteed access to opulent faunal and plant resources.

Alluvial Valley

Gallery forest occupied many of southern California's coastal stream bottoms down to the point where they met the embayments (True 1966:59). Above the wetland zone, leading into the foothills, the river was lined with riparian vegetation. Above the alluvium, the outlying knolls and ridges were covered with a chaparral community (Affinis 1991). While this vegetation probably included a variety of economic species, the zone was probably not as lush as the marsh area below or the inland valley/mountain zones of the middle and upper basin. The coastal live oak, quite abundant in this area, supplied the historic Luiseno with acorns (Bean and Shipek 1978).

The Middle Basin

The middle basin of the San Luis Rey River covers the 57 km reach from Bonsall to Lake Henshaw. Upstream from Pala, the river gradient increases dramatically: from a 3 m/km to a 50-60 m/km average. The middle basin is part of the intermediate foothill and mountain zone of the Peninsular Range (see Figure 37). The Peninsular Range is between 50 km and 80 km wide and contains a number of significant topographical and physiographic features. A wide range of potential subsistence resources is associated with these features (True 1966; True et al. 1974). Physiographically, this province consists of a series of uplifted batholithic plutons that have been dissected by a number of generally westward-flowing streams, of which the San Luis Rey River is one. The eastern margin of the province is marked by a steep and clearly defined scarp separating the interior mountain region from the desert. In contrast, the western face slopes gradually down toward the Pacific Ocean. In north-central San Diego County the Peninsular Range is dominated by the Agua Tibia Mountains, with elevations ranging between 500 and 6,000 feet. Palomar Mountain forms the southernmost anchor of the Agua Tibias. Along the southwestern flank of Palomar Mountain, at an elevation of 2,888 feet, the scarp is broken by a series of flats that mark the location of the Elsinore fault line. At this point, the fault trends southeast to northwest from the Valle de Jose region (presently occupied by Lake Henshaw) towards the town of Pala near the northwestern end of the main Agua Tibia range. The San Luis Rey River follows the margins of the Elsinore Fault zone from Lake Henshaw to Pala, a distance of 38 km, after which it turns southwest for another 37 km to the ocean. From the Agua Tibia-Palomar Mountain block, a large number of tributaries drain southwest into the river.

Many of the prehistoric sites in the middle basin are concentrated on the flats marking the fault zone (True 1978). Springs usually associated with a faulted topography provide water sources for some of these settlements. Vegetation on the flats differs from that of the adjacent terrain and is essentially oak parkland (savannah). Due to differences in elevation, exposure, and parent rock, the terrain adjacent to the parklands supports a wide variety of vegetation. Typical southern California chaparral is established on the steeper mountain sides and canyons along the southwestern boundary of the fault zone. The flanks of Palomar Mountain above the flats support chaparral on the lower slopes and mixed broadleaf coniferous forests higher on the mountain. Small flats or benches on the mountain flanks are marked by oak groves. The canyons contain gallery forest, oak groves and some conifers as low as 3,000 feet elevation. Relatively high orographic precipitation along the southwestern rim of the mountain proper (up to 50 inches a year) allows a dense and quite lush mixed broadleaf and coniferous forest environment at elevations above 5,000 feet (True et al. 1974). This contrasts with rainfall in the lower elevations of less than 15 inches per year.

The Valley Center plateau, although quite a distance from the river channel, is also part of the San Luis Rey River Basin. Located south and west of the river, it is drained by the Key Canyon and Moosa Canyon Creeks, both tributaries of the San Luis Rey River. The plateau consists of a series of broad flats or erosion surfaces with elevations ranging between 1,800 and 2,000 ft. Vegetation is dominated by chaparral, with stream channels hosting gallery forest of oak, willow, cottonwood, and sycamore, in addition to secondary growth including many economic species (True 1966).

The Upper Basin

The upper basin of the San Luis Rey River includes the Lake Henshaw basin and the Eastern Peripheral areas. Lake Henshaw, completed in 1922, controls a drainage area of 531 sq km. Rising just below the crest of Bucksnort Mountain (1,888 m AMSL), the river flows 25 km southeast into the lake area, where it is joined by several other headwater tributaries including the West Fork which rises below Mount Palomar (1,872 m). Before construction of the dam, the Lake Henshaw basin was a relatively flat and wide valley dissected by the tributaries mentioned above. Elevation of the basin, the Valle de San Jose (Portilla), ranges between 2,850 and 2,900 ft. To the west and south, the valley is flanked by steep and clearly defined scarps of the Agua Tibia Mountain range, while to the north the relatively gentle slopes of Bucksnort and Hot Springs Mountain occur. To the east, a series of valleys lead to the desert terrain of the Borrego Valley.

Summary

The San Luis Rey River basin is characterized by a wide range of ecological zones. Clearly, a broad and generally rich subsistence base was at the disposal of the area's inhabitants. The lower basin, with its rich estuaries, lush freshwater marshland, and forested alluvial valley, all surrounded by gently rolling hills covered with chaparral, contained a myriad of readily available coastal and terrestrial resources. The middle basin was similarly blessed with a lavish and diverse resource base. Here, a varied vegetation extended through a number of environmental provinces, each associated with highly localized environmental conditions. As a result, the zone included a wide range of vegetative and animal resources under climatic conditions extremely favorable for aboriginal subsistence on a seasonal basis. The upper basin, although lacking the dramatic ecological variation found downstream, mirrored at least some of the favorable aspects of the middle basin: a fairly lush, stream-dissected valley bottom, and spring-rich mountain slopes. Taken as a whole, and especially during the earlier part of the Holocene, the San Luis Rey River basin had sufficient potential to provide a perennial, year-round subsistence base.

Settlement Patterns and Intrabasin Dynamics

In the following discussion, settlement patterns in the three subdivisions of the basin are identified and linked to potential resources in these areas. Sites are compared, both within each zone and between the various zones. Similarities between sites are viewed from the perspective of coinciding adaptation strategies in which they have a cultural, if not temporal, correspondence. The data are then integrated to determine if the basin can be viewed as one economic entity, with groups of people moving up and down the river during their seasonal collecting rounds. Most of the discussion focuses on the La Jolla/Pauma period represented by the Whelan Lake site. Some thought is given to the possible presence of a San Dieguito substratum in the area. The later occupation of the basin by San Luis Rey I/II groups and the historic Luiseno is also considered. These later archaeological complexes and ethnographic sources provide the necessary context for evaluation of settlement changes through time, while furnishing comparative information on subsistence strategies.

Lower Basin Settlement

A substantial portion of the lower San Luis Rey River basin has been surveyed by archaeologists. Currently, the combined site files of the San Diego Museum of Man and the San Diego State University contain 68 prehistoric and protohistoric sites recorded along the 15 km stretch from the river mouth to just above Guajome Lake (Figure 39, Table 47). When assessing this inventory, it is important to keep in mind that an undetermined number of early sites along the California coast have been inundated. Furthermore, inundation has not been restricted to littoral sites. In major river valleys, sites have been flooded or covered by thick layers of sediments (Inman 1983). Yet, even with these post-depositional biases affecting the archaeological record, an adequate percentage of the lower basin has been investigated to allow robust pattern of settlement to emerge.

Table 47 lists all 68 recorded sites, along with geomorphic context, site characteristics, and known cultural affiliation. To form a convenient framework for examining settlement patterns, the sites have been grouped in five broad areal clusters (Areas A-D). Essentially, the table represents a record of archaeological reconnaissance along the river, starting at the coast and moving upstream to the east, with a brief excursion along Pilgrim Creek, and ending up just west of Guajome Lake.

In the following, settlement along the lower river is reviewed. First, the site inventory is categorized according to site location, composition, function, and age. These four categories are then correlated. Establishing when and why specific sites occurred in a certain place, is the first step in assembling a model of settlement for the entire basin.

Site Location

For much of the San Luis Rey River basin the 60 ft contour line forms the rough boundary between Late Pleistocene river terrace and Holocene alluvium (see Figure 39). This boundary not only extends along the San Luis Rey River, but also for at least 1 km up the river's major tributary, Pilgrim Creek. Tertiary bedrock rises above the Late Pleistocene river terrace, except for the lower 4 km of the basin. Here, it is found directly above the alluviated valley, which tapers increasingly toward the river's mouth. Along the coastline, 100 ft to 350 ft high sandy beach ridges and Pleistocene marine terrace overlie local Tertiary bedrock.

Archaeological sites have been found on four geomorphic surfaces ([#] = number of sites):

(a) Pleistocene marine terrace or beach ridge [1],
(b) Holocene alluvium [12],
(c) Late Pleistocene river terrace [40],
(d) Dissected Tertiary bedrock [19].

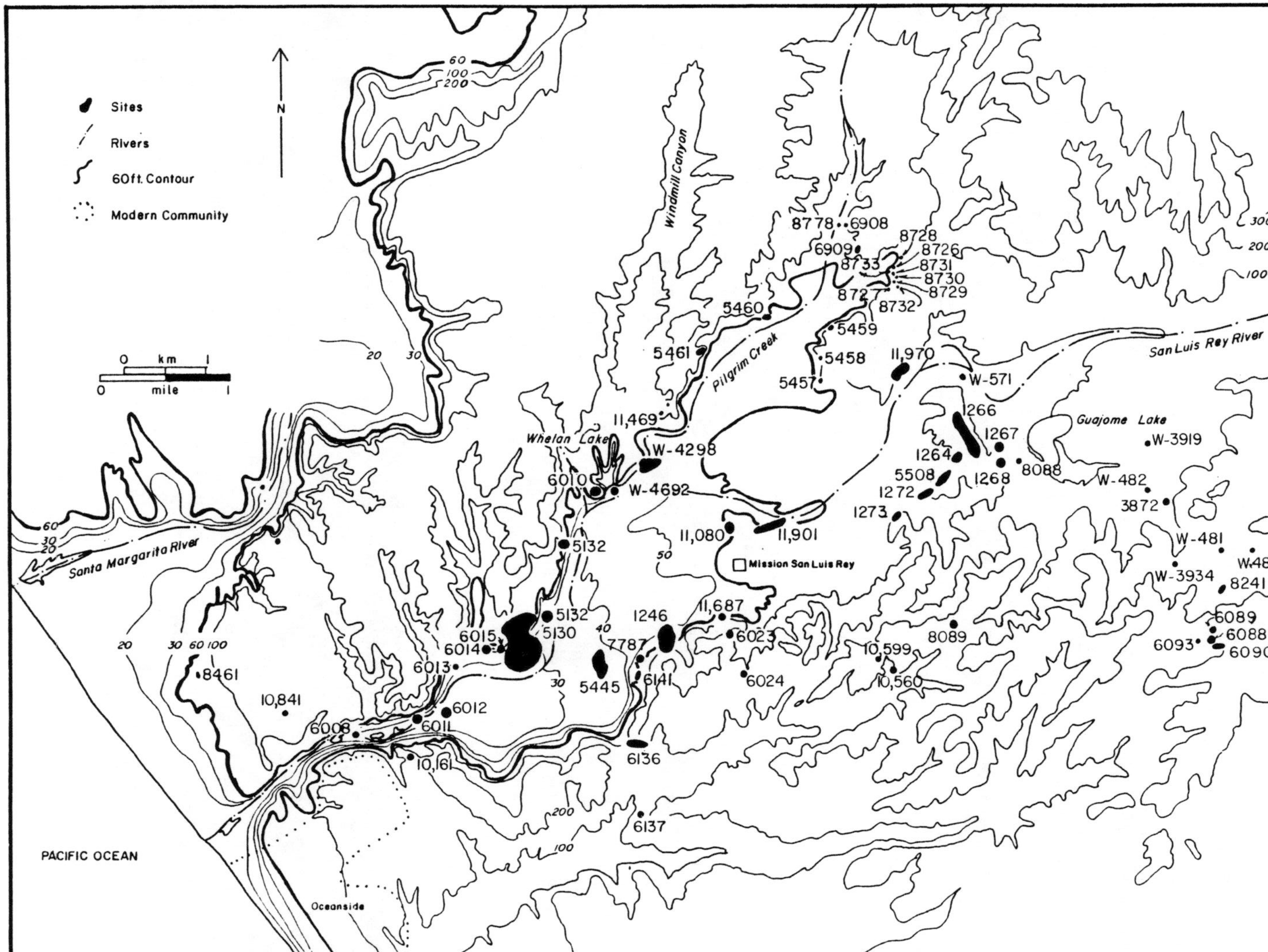

Figure 39. Map of the lower San Luis Rey River basin with recorded prehistoric sites.

Table 47. Prehistoric Sites Recorded in the Lower San Luis Rey River Basin.

Key:

* San Diego State University site designation
** San Diego Museum of Man site designation
*** Ind: Indeterminate
SLR: San Luis Rey I/II period

SDI# *	W# **	Comments	Affiliation***
Area A: North of the San Luis Rey River, south of Whelan Lake.			
10,841	3898	Shell scatter only (60 *Donax*, 12 *Chione* frags.).	Ind
6008	1832	Shell midden, Tizon brown ware.	SLR
6011	1835	Light shell scatter, all *Donax*, brown ware sherds.	SLR
6013	1837	Light shell midden (mostly *Donax*, some *Ostrea* and *Argopecten*	Ind
6014	1838	Fairly dense shell midden (mostly *Donax*, some *Ostrea* and *Argopecten*; ground stone; dated obsidian.	SLR
6015	1839	Shell scatter on former riverbank terrace: mostly *Donax*.	SLR
5130	1361	Large shell midden.	La Jolla/SLR
5132	1363	Shell midden, partly under alluvium, occasionally flooded.	SLR
6009	1833	Light shell (mostly *Chione*) and lithic scatter.	La Jolla?
6010	1834	Shell midden: the Whelan Lake site.	La Jolla/SLR
	4692	Light shell and lithic scatter	SLR
Area B: South of the San Luis Rey River, south of the Mission.			
10,161	3573	Ground stone and lithic scatter, tools with heavy patina	Pauma
6012	1836	Sparse shell midden, mostly *Donax*, no lithics or sherds.	SLR
5445	1527	Broad, but light shell scatter, mostly *Donax*.	SLR
6136	1778	Shell and lithic scatter, including ground stone.	Ind
6137	1779	Shell midden with 1000+ *Donax*, only one *Chione* observed; FCR.	Ind?
6141	1806	Shell midden, mostly *Donax*, some *Chione*, *Argopecten*, and *Ostrea*; sherd; flake.	SLR
7787	3238	Some shell, FCR, lithics, and ground stone.	Ind
1246	3829	Shell scatter, mostly *Chione* and *Argopecten*; burned faunal bone, FCR, lithics, ground stone, three discoidals.	La Jolla
6023	1852	Lithic/ground stone scatter (tested).	Ind
6024	1851	Lithic/ground stone scatter (tested).	Ind
11687		Lithics, ground stone, sherds, shell, faunal.	SLR
10,559	3830	Lithic and ground stone scatter.	Ind
10,560	3830	Lithic and ground stone scatter.	Ind
8089	2232	Light lithic scatter (heavy patina on tools).	Pauma

Table 47. Prehistoric Sites Recorded in the Lower San Luis Rey River Basin (Continued).

SDI# *	W# **	Comments	Affiliation***
Area C: Along the San Luis Rey River, above the Mission.			
10,080	3549	Shell scatter, mostly *Donax* with some *Argopecten* and *Chione*, debitage, proj. point.	SLR
11,961	4592	Light shell scatter (*Chione* and *Argopecten*), Ground stone, lithics, Tizon brown ware, faunal bone.	SLR
1273		Sherds and mano: no shell	SLR
1266	569	Large and dense milling stone assemblage, including one discoidal, two sherds, and one shell.	Pauma/SLR
1264		Ground stone only (heavy patina on tools)	Pauma
1267		Same	Pauma
1268		Same	Pauma
1272		Manos, scrapers, hammer stones (patinated)	Pauma
5508	2422	Hammer stones, debitage, FCR, some shell	Pauma
8088	2231	Manos, chopper, scraper plane, flakes (some patina)	Pauma
	571	Bedrock mortar	SLR
11,970	4600	Bedrock mortars, lithics, ground stone, sherds.	SLR
Area D: Along the San Luis Rey River, east of Guajome Lake.			
	3919	Bedrock mortars	SLR
	482	Bedrock mortars	SLR
8872	568	Bedrock mortars and ground stone.	SLR
	3934	Grinding slicks without artifacts.	SLR
8241	479	Bedrock mortars, lithics, sherds.	SLR
8242	480	Bedrock mortars, pictographs, lithics, sherds, shell.	SLR
	481	Bedrock mortars, manos, sherds, lithics.	SLR
6088	1918(A)	Bedrock mortars, two shells, burned faunal bone, lithics.	SLR
6089	1918(B)	Ground stone, hammerstone with heavy patina, some chipped stone.	Pauma
6090	1918(C)	Bedrock mortars , lithics, no ecofacts.	SLR
6093	565	Bedrock mortars and slicks, lithics, no ecofacts.	SLR
Area E: Along Pilgrim Creek			
	4298	Light shell and lithic scatter.	Ind
11,469	4299	Five bedrock mortars, no artifacts or ecofacts.	SLR
5461	1529(E)	Broad but light shell scatter (about 200 *Chione* and 10 *Argopecten* fragments), one mano fragment.	La Jolla?
5460	1529(D)	About 100 *Chione* fragments on the surface, one flake.	La Jolla?
5459	1529(C)	Extensive but light shell scatter (50 % *Chione*, 50 % *Argopecten*); one flake.	La Jolla?
5458	1529(B)	About 10 *Chione* fragments observed.	La Jolla?
5457	1529(A)	Extensive but sparse shell scatter (about 50 *Chione*., and several fossilized *Ostrea* fragment observed); lithics.	La Jolla?

Table 47. Prehistoric Sites Recorded in the Lower San Luis Rey River Basin (Continued).

SDI# *	W# **	Comments	Affiliation***
8778	2226(B)	Milling features only, no artifacts.	SLR
6908	2226(A)	Bedrock mortars, ground stone, pictographs, sherds.	SLR
6909	2227	Large camp with abundance of milling stones.	Pauma
8726	2738	Bedrock mortars and slicks.	SLR
8727	2739	FCR, lithics, sherds.	SLR
8728	2740	Food processing camp (obsidian date of A.D. 1515).	SLR
8729	2741	Pictographs	SLR
8730	2795	Single slick	SLR
8731	2796	Some *Donax*, a sherd, lithics	SLR
8732	2797	Slick	SLR
8733	2798	Mortar	SLR

(a) Pleistocene Marine Terrace/Beach Ridge. No prehistoric sites are known on the open coast around the mouth of the San Luis Rey River (Cupples 1976; Cupples and Hedges 1977; Jones 1991:Figure 3). Both sea level rise and modern development may be to blame for this void. A more direct cause, however, may be the relative harsh and unprotective nature of the littoral environment. Fog, wind, cold, in addition to the lack of potable water, probably all conspired in repelling aspiring settlers. In the lower basin, the closest site (SDI-10,161) to the coast is recorded 1.5 km inland. This is the only site occurring on the Pleistocene marine terrace.

(b) Holocene Alluvium. Twelve sites have been found on Holocene alluvium in the floodplain of the river. Much of this alluvium was deposited long after 5000 B.P., and it is obvious that these sites are relatively recent. A substantial number of early sites may have been submerged under alluvium.

(c) Late Pleistocene River Terrace. River terraces were the favored site location. Sites situated in this context were safe from river floodings and alluviation, but at the same time close to fresh water and, in some cases, to the estuarine zone.

(d) Dissected Tertiary Bedrock. Sites on Tertiary bedrock are located at somewhat higher elevations and further inland compared to the other components. Mostly located along small tributaries, they are some distance away from the main river channel.

Site Composition

The 68 lower basin sites have been grouped into six categories based on differences in site composition. These categories are as follows ([#] = number of sites):

(a) Shell midden/scatter including artifacts [20]
(b) Shell midden/scatter without artifacts [7]
(c) Artifact scatter with some shell [2]
(d) Artifact scatter with no shell [19]
(e) Millingstone features with/without artifacts [18]
(f) Pictographs [3]

(a) Shell Middens/Scatters with Artifacts. Of the six categories, Class (a) sites are the most substantial in terms of average size and ecofact/artifact mass. The most extreme example is SDI-5130, a large (240 m by 200 m) multicomponent, village site consisting of several deep (circa 1 m) midden areas, including hearths, storage pits, and burial and cremation features. Although considerably smaller than SDI-5130, the Whelan Lake site (SDI-6010) also has deep deposits and a number of hearths. No excavation results have been published for other sites in this class, and for many insufficient data are available to determine if the site is an actual midden or just a surface scatter. Locations for shellfish-dominated sites vary, but the majority (9) occur along the north bank of the San Luis Rey River, below Whelan Lake. All occur either on low knolls formed from the Late Pleistocene river terrace or on Holocene alluvium. None have been found further inland on Tertiary bedrock.

(b) Shell Scatters Without Artifacts. With seven sites, the Class (b) sample is relatively small. The inventory includes four shell scatters along Pilgrim Creek, each of which contained only one artifact. Location of Class (b) sites is either close to the coast (SDI-10,841, SDI-6012, and SDI-6013), or along Pilgrim Creek (SDI-5458/5461). The sites occur either on the Late Pleistocene river terrace or on Holocene alluvium.

(c) Artifact Scatters With Some Shell. Class (c) sites are sparsely distributed. All three known examples of this class (SDI-1266, SDI-5508, and SDI-11,687) are located on the Late Pleistocene river terrace, and for each site not more than a couple of shell fragments were recorded.

(d) Artifact Scatters Without Shell. Class (d) sites are relatively common. Twelve of the 19 sites are found on the Late Pleistocene terrace and seven on Tertiary bedrock; none have been found on alluvium. Nor have any been recorded west of the San Luis Rey River. Nine sites are dominated by milling stones and lithic tools with heavy patina. Composition of others varies, often including lithic debitage and/or sherds.

(e) Milling Features. Class (e) is formed by sites with bedrock mortars or slicks. None of these sites occur closer to the coast than 8 km. Two major clusters occur on the Late Pleistocene river terrace along Pilgrim Creek and on bedrock east of Guajome Lake. An anomaly is W-571, a single bedrock mortar adjacent to the river channel in an otherwise alluviated area. Whereas a number of the sites consist of single or groups of milling features with no artifacts, others have associated manos, pestles, lithic tools, debitage, or a few sherds or shell.

(f) Pictographs. Three sites (SDI-6908, SDI-8729, and SDI-8242) recorded in the lower basin have been grouped in Class (f). Along Pilgrim Creek, two groups of pictographs were recorded, one associated with bedrock mortars and artifacts. East of Guajome Lake, a site with bedrock mortars and artifacts contains pictographs.

Site Function

Site composition generally provides the basis for inferences about site function. Comparison of functional site classes with the preceding categorization based on site composition demonstrates the overall congruity between the two categories. The following site function categories are identified in the lower basin inventory (Figure 40) ([#] = number of sites):

(a) Camp associated with a mixed marine and terrestrial subsistence economy [19]
(b) Shellfish procurement [8]
(c) Camp associated with terrestrial resources [26]
(d) Plant milling station [12]
(e) Ceremonial [3]

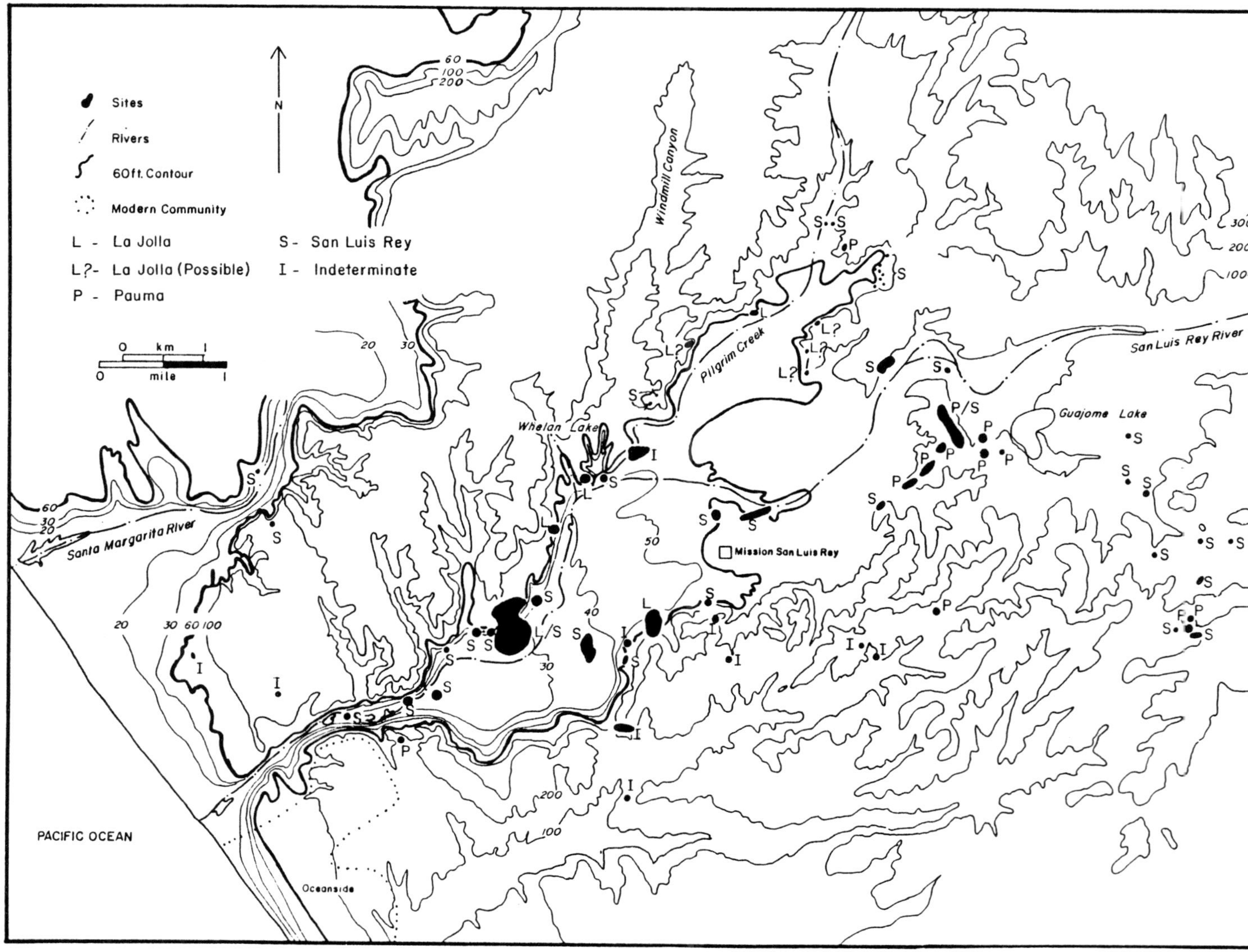

Figure 40. Lower basin sites identified by cultural affiliation.

(a) Campsites Associated with a Mixed Marine/Terrestrial Subsistence Base. The 20 shell middens/scatters with artifacts are all interpreted as camp sites whose occupants procured and processed shellfish, fish, small mammals, and plants. Size of these camps varies. Most are fairly small surface scatters, reflecting limited and highly seasonal use; others have thick midden deposits (e.g., SDI-5130, SDI-6010, SDI-1246), indicating use over a longer period of time and possibly a degree of permanence. Many of the more substantial middens and scatters have subsurface hearths or clusters of fire-cracked rock on the surface. Besides shell, ecofacts include varying densities of animal bone. Chipped stone tools and milling stones further exemplify an economy balancing a diverse subsistence base.

(b) Shellfish Procurement. All sites consisting of shell with no artifacts or sites of shell with only one or two artifacts, are interpreted as procurement locations for shellfish and possibly other marine food. Rather than a place where groups would stay for an extended period, these minor scatters connote locations where shellfish were cleaned and then either consumed "on the spot" or transported further inland. It is likely that many such sites were located close to the ancient estuary and most have probably been submerged under alluvium. The placement of the known examples of this functional class, either relatively close to the coast or fairly far inland is enlightening. All four coastal scatters are dominated by *Donax*, a shellfish genus favoring open-coast sandy beaches (Ricketts et al. 1985). The coastal examples include the only site (SDI-10,841) in the lower basin site inventory that is situated on marine terrace. Another one of these sites (SDI-5445) is a substantial midden located on alluvium in the floodplain. The four inland procurement sites are all located away from the main river channel, along Pilgrim Creek. Artifacts include no more than a sporadic flake or ground stone fragment. The inland sites are dominated by *Chione* and *Argopecten*, all estuarine species. There is no evidence that the estuary would have extended up as far as Pilgrim Creek. The shellfish, then, would have been gathered further downstream and brought to these sites, which would have been situated at the edge of the marsh. These sites then may represent activities of inland groups who cleaned or ate their catch from a coastal expedition.

(c) Campsites Associated with Terrestrial Resources. Camps whose occupants targeted terrestrial resources form the largest functional class. Both artifact scatters with little or no shell, and milling feature sites with artifacts are included in the group. Milling stones dominate most of the assemblages. Like the mixed economy camps, these terrestrial resource procurement and processing camps vary in size. Some of the larger examples are SDI-1266, SDI-6909, and SDI-6088. SDI-1266, located on a river terrace overlooking the San Luis Rey River to the north, is described as a large, multicomponent, milling stone assemblage, including discoidals and sherds. The site is surrounded by a number of smaller workshops and milling stations. SDI-6909 (along Pilgrim Creek) and SDI-6088 (away from the river, west of Guajome Lake) are both fairly large camps including milling stones and lithic tools displaying heavy patina.

(d) Plant Milling Stations. Sites consisting solely of milling features (bedrock mortars or slicks) or comprised of milling features and ground stone artifacts, are classified as plant milling stations. Such sites are often associated with neighboring campsites.

(e) Ceremonial. Of the three groups of pictographs identified in the lower basin, one occurs as an isolated site (SDI-8729) without artifacts, whereas the others are part of milling stations. In absence of a more specific designation, pictograph components are classified as ceremonial.

Cultural Affiliation

For a number of sites, diagnostic artifacts or specific features are used as proxies for cultural affiliation. For others, age has been secured through absolute dating methods. Age of an additional group of components can tentatively be established by (1) evaluation of patina present on lithics, (2) predominance of *Chione/Argopecten* or *Donax* in shell assemblages, or (3) location of sites on Holocene alluvium. For some of these sites, the culture designation has been prefixed with the qualifier "probable". For the remainder, cultural definition could not be established. Figure 40 is a map of the lower San Luis Rey River basin with sites labeled according to cultural affiliation. The following classes have been identified ([#] = number of sites):

(a) La Jolla [3]
(b) Probable La Jolla [6]
(c) Pauma [11]
(d) San Luis Rey (I/II) [39]
(e) Indeterminate [9]

A brief explanation of some of these terms is necessary. La Jolla and Pauma are best viewed as functional variants of the same archaeological complex (see Chapter 2). The difference is primarily locational, and to a lesser degree artifactual. La Jollan subsistence focused on the resources available in lagoonal environments: primarily fish, shellfish, water fowl, small land mammals, and local plants (Erlandson 1991; Gallegos 1991; Gibson et al. 1990). Not surprisingly, La Jollan sites are located near the coast. In contrast, Pauma sites occur inland. Assemblages of La Jollan sites reflect a mixed marine and terrestrial adaptation, whereas Pauma sites demonstrate an exclusively terrestrial subsistence economy. Absolute dates from Pauma sites are rare, but the two complexes appear to be contemporaneous.

As discussed above, the lower basin inventory includes a functional class reflecting a mixed marine/terrestrial economy and one corresponding to an unmixed terrestrial adaptation. Examining site composition, it is striking how little overlap occurs between these two functional classes; sites are either dominated by shell or, conversely, only an occasional shell occurs. One would expect a gradation of the one class into the other, but, apparently there is no middle ground. Absolute dates and diagnostic artifacts suggest that a number of sites from the two classes are contemporaneous and used by La Jolla/Pauma groups. Thus, we are faced with the rather perplexing phenomenon of areal coexistence of two adaptational distinct site types that presumably represent the same archaeological culture. Instead of labeling both site types La Jollan, it seems more appropriate and convenient to follow the traditional distinction based on artifactual traits. Thus, the early shell midden camps are designated La Jollan, whereas the contemporaneous milling camps are labeled Pauma, even though this latter group occurs outside the Pauma "culture" area.

<u>(a) La Jolla</u>. Only three unequivocal La Jollan sites have been identified: SDI-5130, SDI-6010, and SDI-1246, all substantial shell middens located on knolls overlooking the San Luis Rey River. The ages of the first two of these have been determined through radiocarbon dates obtained from shell samples. SDI-6010, the Whelan Lake site, was probably occupied between 7,400 and 6,500 B.P. SDI-5130 has two components but was primarily occupied between 6,300 and 4,000 B.P. For SDI-1246 La Jollan affiliation is indicated by diagnostic artifacts (e.g., discoidals) on the surface. At all three middens shell is dominated by *Chione*, followed by *Argopecten*.

<u>(b) Probable La Jolla</u>. Six sites have the potential to be La Jollan, but no conclusive evidence is available. One of these is a light shell and lithic scatter, located on a knoll halfway between SDI-6010 and SDI-5130. The others are shell scatters along Pilgrim Creek. All are located on the Late Pleistocene river terrace and all are dominated by *Chione* and *Argopecten*. These two shell species, especially *Chione*, prefer an estuarine habitat. As discussed above, the lower portion of the river probably formed a protected, brackish lagoon between 8,000 and 4,000 B.P. It is likely that most of the *Chione* and *Argopecten* species occurring at archaeological sites in the lower basin date to this estuarine period. All three sites in the lower basin positively identified as La Jollan have shellfish assemblages dominated by these species. The same is true for all known "lagoonal" La Jolla sites. Later sites in the

lower basin, for instance, those situated on Holocene alluvium, are characterized by low percentages of these estuarine clams and scallops. In sum, the probability is high that the six *Chione*-dominated shell scatters in the lower basin are La Jollan.

(c) Pauma. The Pauma complex is typified by artifact scatters consisting of basin metates, manos, occasional scraper planes, and, rarely, discoidals. Bedrock mortars, pictographs, ceramics, and small triangular projectile points are absent at Pauma components (True and Beemer 1982). Lower basin sites were designated as Pauma if little or no shell was observed and the assemblage was distinguished by a high proportion of milling stones, with tools often displaying patina. Pauma sites were also recognized by the presence of discoidals, the most distinct diagnostic artifacts of this complex. Eleven lower basin sites have been identified as Pauman. Two of these, SDI-1266 (along the river, east of the Mission) and SDI-6909 (along Pilgrim Creek) are fairly large and dense milling stone scatters. SDI-1266 also contains a discoidal on the surface. Patina marks many of the artifacts. The remaining nine Pauma sites are smaller artifact scatters, mostly dominated by ground stone and containing highly patinated tools and lithics. Seven of these sites are clustered around the large Pauma camp, SDI-1266, for which they may have served as milling stations and workshops. Site location varies. All Pauma sites are located somewhat inland, above 100 ft elevation. The placement of SDI-10,161, relatively close (2.7 km) to the coast and overlooking the Oceanside Narrows, is intriguing. Artifacts consist of patinated scrapers, core tools, and flakes, without ground stone or shell. Situated on the edge of marine terrace and bedrock, the coastal scrub vegetation is dominated by yucca, suggesting that the site may have been a procurement camp for this species.

(d) San Luis Rey. With 39 components, San Luis Rey I/II (SLR) forms the largest cultural site class. All sites containing bedrock mortars or slicks, pictographs, sherds, or small triangular-shaped projectile points, are designated SLR. In addition, all sites on recent alluvium were placed in this class. Fourteen of the SLR sites are shell middens or scatters, 13 are camps targeting terrestrial resources, 11 are plant milling stations, and three sites contain pictographs. At a number of the larger sites (e.g., SDI-1266, SDI-5130, and SDI-6010), a small San Luis Rey occupation is found on top of the more extensive La Jolla or Pauma occupation. For the SLR complex no terminological distinction has been made between sites emphasizing coastal subsistence and those focusing on terrestrial resources. In many ways this complex reflects the same adaptational pattern noted for the La Jolla/Pauma. Sites are associated either with a mixed marine/terrestrial or with an unmixed terrestrial economy. Some gradation between these two functional categories is noticeable, however. Several of the larger SLR plant processing camps have some shell, whereas at several of the shellfish procurement/processing camps a relatively high density of lithic material was noted. All SLR shell assemblages are dominated by *Donax*, instead of *Chione* as earlier. This change may reflect a deterioration of the estuarine habitat due to silting of the lagoon.

Discussion

Settlement in the San Luis Rey River basin did not occur in a vacuum. Along the coast of southern California a number of rivers of various sizes drain into the Pacific. Many of these have associated lagoons, some of which are still estuarine. Most early coastal sites are situated on the perimeter of what once were fjord-like embayments formed by marine transgression over deeply incised river valleys. It has been suggested that the initial occupants of the coastal area migrated from the Great Basin where lacustrine environments dried up at about the same time (ca. 9,000 B.P.) that the lagoons were formed (Gallegos 1991). The similarity of coastal estuaries to the pluvial lakes may have allowed the newcomers to adapt quickly to the coastal resource-rich habitats. A substratum of San Dieguito-like artifacts occurs at both La Jollan and Pauma sites (e.g., Gallegos 1987, 1991; True 1978) and the idea of a San Dieguito-La Jolla/Pauma transition, though still heavily debated, is plausible.

The above inquiry into settlement along the lower drainage of the San Luis Rey River identified a number of correlations between site location and site function. Four basic functional site classes were defined: (1) processing camps focused on shellfish, (2) shellfish cleaning/consumption stations, (3) processing camps focused on plants, and (4) plant milling stations. The earlier, La Jolla, examples of Class (1) are all located on knolls on the Pleistocene river terrace, between 4.5 km and 7 km from the coast. Later, San Luis Rey shell processing camps occur between 2.3 km and 12 km from the coast, and are found not just on the river terrace, but also on alluvium and bedrock. Shell cleaning or consummation stations are found either relatively close to the coast or further inland along Pilgrim Creek, a tributary to the main river channel. Whereas La Jolla ecofact assemblages are dominated by *Chione* and *Argopecten*, San Luis Rey components contain primarily *Donax* species. These differences in site composition appear to reflect ecological changes through time. Plant procurement/processing camps are found on both the river terrace and bedrock, but never on alluvium. Interestingly, none occur along the west bank of the lower San Luis Rey River. This area, enclosed between the San Luis Rey and the Santa Margarita rivers to the north, appears reserved for camps associated with a coastal economy. A major difference between the earlier Pauma and later San Luis Rey sites is that bedrock mortars are restricted to the San Luis Rey sites. This may indicate an intensified emphasis on hard seed (e.g., acorn) grinding by San Luis Rey groups.

For the La Jolla period, several lines of evidence suggest that an estuary was present in the lower basin. First, all La Jollan shell assemblages are dominated by *Chione* and *Argopecten*, both estuarine species. *Chione*, a genus favoring the sand and mud flats of of shallow water, is dominant, suggesting that a major portion of the lagoon was composed of this habitat. This notion is contrary to the geomorphological assessment of estuarine potential for the lower basin. If (as concluded in Chapter 2) the lagoon was confined to the river mouth and the 2.5 km long deeply incised Oceanside Narrows, shallow water and mud flats would have been rare in much of the lower basin.

The location of the La Jollan sites forms a second line of evidence arguing for an estuary extending further inland than the Oceanside Narrows. The three large La Jollan sites SDI-6010, SDI-5130, and SDI-1246, surround the broad, flat, and apparently heavily alluviated portion of the lower basin above the Narrows. SDI-6010, located furthest inland, was occupied between 7,400 and 6,500 B.P., and SDI-5130 from about 6,300 to 4,000 B.P. If, as assumed, these sites were located optimally so as to have well-balanced access to freshwater and estuarine resources, the temporal difference between the two locations may reflect an ecological change, namely the increasing silting of the lagoon, forcing the shellfish collectors to move their camp downstream. Maybe, when the Whelan Lake site was occupied, the SDI-5130 location was adjacent to the lagoon and too far from fresh water to be chosen as a camp site. Only after the estuary had receded and freshwater marshland began to form around the river channel, was SDI-5130 established. In this scenario, the estuary may have encompassed a portion of the broader, lower valley. This sheltered lagoon would have had shallow water, while largely consisting of mudflats. This reconstruction of events remains highly speculative, of course, and only further geomorphological work in the area will settle the issue.

Whatever the explanation, the location of the Whelan Lake site, on an elevated knoll near a fresh water source, conforms to the pattern noted for most early sites on the California coast. These elevated, early site locations have been explained in various ways: proximity to ancient oak forests (Rogers 1929:343); defensive needs combined with the more terrestrial focus of early economies (King 1981:150); optimal placement to balance the marine and terrestrial aspects of a mixed economy (Moore 1985); and reflecting the instability in canyon and valley bottoms due to submergence, erosion, and sedimentation (Erlandson 1988). The Whelan Lake site occupies a unique place in the lower basin. The La Jollan shell midden is located at the juncture of Pilgrim Creek and the San Luis Rey River, and thus at a strategic point between the two areas of shell distribution noted in our overview. Several more minor creeks enter the river close to the site, and where Whelan Lake currently stands there was probably a pond or freshwater marsh. More than likely, this location was very desirable, representing an optimal balance between estuarine, fresh water, and terrestrial resources. In addition, the site's occupants were able to monitor movement of people and goods between the coast and upstream on the San Luis Rey River and Pilgrim Creek. Possibly, the site was a major hub of early La Jollan activities in the lower San Luis Rey River basin. In this respect, its location some distance from the ancient lagoon is no anomaly.

The end of the La Jolla occupation coincides with increased alluviation of the estuary, and, as elsewhere on the coast, the archaeological record becomes silent. Not until A.D. 1400, with the advent of the San Luis Rey sequence, is large scale reoccupation of the area apparent. The larger shell middens and plant processing camps of the La Jolla/Pauma period are being reused, and a large number of new sites appear. Several shell middens appear on alluvium in the sheltered lower valley. A changed environment is apparent from the dominant species in these new middens: *Donax*, a genus found along sandy beaches of the open coast. The plant processing sites also contain a new element, bedrock mortars, implying an intensified use of hard seeds such as acorns. To assess the dynamics involved in the repopulation or the more intensified use of the lower basin, events occurring further upstream need to be examined.

Middle Basin Settlement

The central San Luis Rey drainage, including the major rancheria territories of Pala, Pauma, Cuca, and La Jolla, is one of the most intensively archaeologically surveyed areas in southern California. Although much of the work here focused on the late prehistoric (San Luis Rey I) and protohistoric (San Luis Rey II and Luiseno) periods (True 1966; True et al. 1974), an inland variant of the La Jolla complex, termed Pauma, was defined based on a sample of sites in the Pauma Valley and the Valley Center Plateau (True 1958, 1978). Sporadic San Dieguito-like artifacts have been found at Pauma sites, suggesting a previous San Dieguito presence in the area. While referring to later periods and occasionally borrowing from ethnographic (Luiseno) sources, the following discussion is primarily concerned with the Pauma occupation of the basin.

The Pauma and La Jolla archaeological complexes are similar in many ways. Pauma artifact assemblages are identical to La Jollan assemblages with two notable exceptions. Pauma assemblages are dominated by milling stones, in contrast with the coastal La Jollan assemblages in which scraping/chopping tools are prevalent (True and Beemer 1982). The higher percentage of ground stone indicates that, instead of targeting lagoonal resources, Pauma subsistence depended mostly on terrestrial resources, primarily small land mammals, local plants, acorns, and other hard seeds (True 1958, 1980; True et al. 1974). The Pauma and La Jolla are probably best understood as functional variants of a same cultural manifestation.

Figure 41 depicts Pauma sites recorded in the San Luis Rey River basin. All sites are located on relatively high ground with respect to the more recent San Luis Rey sites. These locations include knolls, saddles, and old terrace-like alluvial formations often associated with mudflow deposits. As a norm, the sites are on "old" soils developed on mudflow and fanglomerate formations believed to be of Pleistocene age. Many of these mudflow formations are directly related to the Elsinore fault.

Pauma sites are generally found in inland valleys and sheltered canyons, beyond the reach of marine resources. These valleys and canyons contain a wide variety of vegetation, including oak woodland, broadleaf coniferous forest, and chaparral. Beginning at around 8,500 B.P. a slow shift to an emphasis on hard seeds, particularly the acorn, is evident in the archaeological record for southern California. A transition towards a balanophagous economy (intense reliance on acorns) is generally assumed to have begun during this period in California (Basgall 1987; Baumhoff 1963; Chartkoff and Chartkoff 1984).

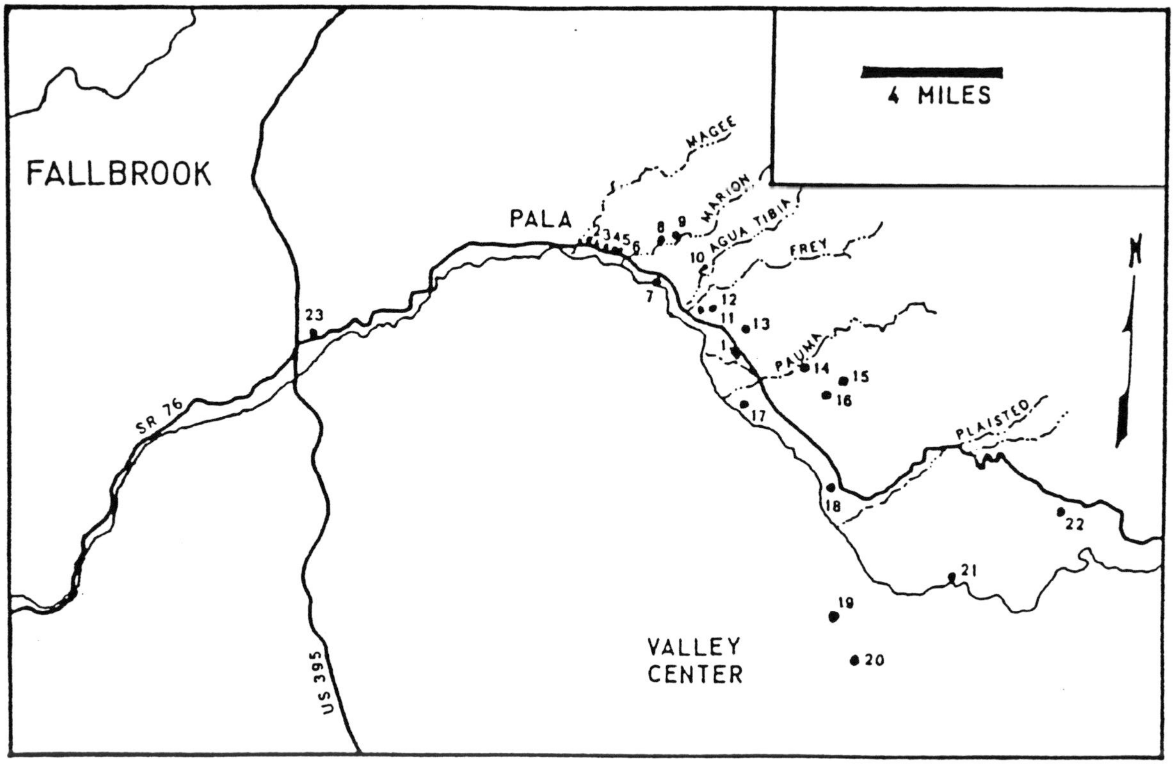

1. SDi-505 (Rincon-47)
2. SDi-716 (Rincon-20)
3. SDi-717 (Rincon-20A)
4. SDi-718 (Rincon-20B)
5. SDi-719 (Rincon-20C)
6. SDi- (Rincon-20D)
7. SDi-513 (Rincon-130)
8. SDi-774 (Rincon-43)
9. SDi-511 (Rincon-43A)
10. . SDi-510 (Rincon-41A)
11. Rincon-300
12. Rincon-301
13. SDi-267 (Rincon-71)
14. Rincon-100
15. SDi-303 (Rincon-133)
16. SDi-512 (Rincon-93)
17. SDi-733 (Rincon-80)
18. SDi-26 (Rincon-129)
19. SDi-346 (Rincon-137)
20. SDi-665 (Rincon135)
21. SDi-16 (Rincon-30)
22. SDi-308 (Rincon-77; Molpa)
23. SDi-682 (Rincon Pala-8; Pankey site)
24. Santa Margarita-1

Figure 41. The middle basin of the San Luis Rey River with recorded Pauma sites (from True 1980).

While there are some 15 species of oak in California, only six are present in significant quantities in the San Luis Rey River Basin. The six species are the scrub oak (*Quercus dumosa*), maul oak (*Q. chrysolepis*), coast live oak (*Q. agrifolia*), interior live oak (*Q. wislizenii*), black oak (*Q. kelloggii*), and Engelmann oak (*Q. engelmanni*). Each species has its associated environmental constraints and preparational problems. Oak productivity varies per species. According to Basgall (1987), oak trees are subject to cyclical productivity independent of the environment, and also vary correspondingly due to environmental fluctuations. Bock and Bock (1974) suggest that areas with multiple oak taxa may provide a larger buffer against total crop failure than areas with only one species.

Information concerning most of the economically important oak species was outlined by Baumhoff (1963). These data are essentially restated here. The scrub oak, relatively scarce in the river basin, generally produces a small crop of bitter tasting acorns. It was not preferred by aboriginal collectors. The maul oak, although large in size, occurs in diffuse populations and produces a good crop no more than once every three years. Consequently, this species was also not preferred by collectors. The coastal live oak, although very abundant in the river valley, is a small tree that produces only small nuts, no more than every other year. The interior live oak, common in the upper river valley, produces a good crop every other year, and was probably the oak most exploited by the Pauma. The black oak is a stable producer of large crops, much revered in ethnographic accounts. The species is confined to small quantities, only growing on the higher slopes. Information on the Engelmann oak, other than that it grows in the same zone as the black oak, is limited, although Luiseno informants claim that it was used for food (True et al. 1974).

Acorn utilization involves four principal steps: harvesting, transporting, storage, and processing. Harvesting is a relatively simple matter, and regardless of the level of technological sophistication would have posed little problem for early Pauma groups. The same groups, however, would have likely encountered some rather troublesome transportation and storage problems. Without an efficient means of transport or effective storage facilities, the early groups were probably limited to exploiting a small area around the village. Processing, the most critical step in the exploitation of acorns, involves shelling, grinding (into meal), and leaching the nuts (with water) (Driver 1953; Gifford 1936). Leaching time varies depending on the species of oak, and whether or not warm water is used (generally three to 12 hours). Preparation is clearly a time-consuming process.

The role of the acorn as a dietary staple for the local populations is well documented in historic times (Baumhoff 1963; Bean and Shipek 1978; Kroeber 1925; White 1963). Determining the importance of the acorn in prehistoric subsistence practices is more difficult, but it is assumed that the crop played a significant role in the diet of the inland groups. Although the cycles of oak species may be predictable, the environmental variability demonstrated earlier and reliance on only two or three species of oak can clearly lead to an inconsistent harvest. Thus, similar to sustenance obtained from estuarine resources, the dietary contribution of acorns would not have been always stable and predictable. The uncertain nature of these forces would require the inland populations to construct social mechanisms to provide a buffer against the environment.

Lower and Middle Basin Comparison

Absolute dates are sparse for the San Luis Rey River basin sites. In the middle basin, only one Pauma component (SDI-682, the Pankey site) has yielded radiocarbon dates. Assays prepared from *Chione* and *Argopectin* samples indicate that the Pauma component was occupied between 6,000 B.P. and 5,000 B.P. (True and Pankey 1985). In the lower basin, only two La Jollan sites are temporally secure: the Whelan Lake site (7,400 to 6,500 B.P.) and SDI-5130 (6,300 to 4,000 B.P.). It is clear that the two complexes are at leasts roughly contemporaneous, although more dates are needed to evaluate La Jolla/Pauma settlement change through time.

Even without more dates, it is possible to examine the relationship between La Jolla and Pauma from a theoretical perspective. A first step in this examination is to study information from ethnographic accounts. This may help fill in some of the lacunae in our knowledge concerning seasonal gathering strategies. From about A.D. 1400 into historic times, most of the drainage of the San Luis Rey River and that of the Santa Margarita River to the north, was inhabited by the Luiseno, who represent the westernmost extension of the Shoshonean subfamily of the Uto-Aztecan linguistic stock (True 1966; Bean and Shipek 1978). Thus, their habitat covered the whole range of ecological zones described for the river basin. The core area of Luiseno occupation coincided roughly with the middle basin of the San Luis Rey River. Villages, each with specific hunting, collecting, and fishing areas, were located in diverse ecological zones. Typically these were in valley bottoms, along streams, or along coastal strands near mountain ranges. Most inland groups also had fishing and gathering sites on the coast that they visited annually when tides were low or when inland foods were scarce from January to March. Each year for the acorn harvest (October-November), most of the village population would settle for several weeks in the mountain groves to collect acorns, hunt game animals, and collect whatever else was available. Most Luiseno foods, however, were available within a day's travel of the village (Bean and Shipek 1978). Acorns were the most important food source, with six species used. Seeds were the next most abundant plant food used, followed by greens, bulbs, roots, and fruits. Depending on location game, fish, and marine animals formed varying percentages of the diet.

For the central drainage of the San Luis Rey River, the highest site densities have been recorded in the interior upland region along the upper central portion of the drainage. For this area a bi-polar subsistence-settlement model has been proposed. This model, largely based on ethnographic evidence, involves seasonal relocation of principal villages or camps along a vertical route following diverse plant procurement zones (True 1966; True et al. 1974; True and Waugh 1982). In the lowland subsistence systems such a vertical diversity in resources was no factor, and no ethnographic or archaeological support for seasonal relocation has been found (True et al. 1991). Compared to the acorn-dominated economies of the middle and upper basin, lowland and intermediate elevation sites appear to have processed a wide variety of vegetal and animal products. In the middle basin, shellfish remains are rare, becoming increasingly more common downstream from Pala, culminating in the shell middens occurring downstream from Whelan Lake.

The lower basin appears to conform to a bilateral subsistence-settlement model, with sites demonstrating either a mixed marine/terrestrial (La Jolla) economy or a terrestrial (Pauma) economy. During both La Jolla/Pauma and San Luis Rey periods, the boundary between these two disparate subsistence strategies was well-defined, although a slight overlap is apparent at some of the San Luis Rey sites. The nature of the relationship between La Jollan and Pauman groups is far from clear and awaits further research.

Archaeological and ethnographic sources provide information on various subsistence resources in the local diet. Table 48 is a comparison of food sources for coastal and inland bands of the historic Luiseno. Table 49 compares frequencies of specific artifact types between Pauma and La Jollan assemblages.

Table 48. Food Sources for the Historic Luiseno (after Bean and Shipek 1978).

	Inland Bands	Coastal Bands
Acorns	25-45%	15-25%
Seeds	20-40%	20-40%
Greens	10-12%	5-10%
Bulbs, roots, fruits	10-13%	10-15%
Game	15-20%	5-10%
Fish and Marine animals	0-5%	20-35%

Table 49. Comparison of Artifact Frequencies Between Pauma and La Jollan Assemblages (after True and Beemer 1982).

Artifact	Pauma	La Jolla
Milling	63.26 %	25.01 %
Scraping, Planing, and Chopping	12.88 %	53.86 %
Hammers	9,73 %	15.86 %
Hammer Grinders	8.51 %	0.30 %
Bifacial	3.89 %	2.39 %
Other	1.70 %	2.58 %

Comparison of the published artifact inventories of Pauman and La Jollan assemblages listed in Table 49 demonstrates that the material remains within the assemblages are essentially identical, the only difference being quantitive in nature. Interestingly, when juxtaposing Tables 48 and 49, it appears that the coastal/inland ratios for various ethnographic food sources are roughly similar to the La Jolla/Pauma ratios for occurrences of artifacts connoting procurement and processing of these subsistence items. The ratios between lithic tool classes in the assemblage of SDI-6010 (Table 46) is roughly similar to the La Jollan ratios of Table 49. It appears, however, that ground stone and bifacial artifacts are more frequent at Whelan Lake than usual for La Jollan sites. Of course, the small size of the Whelan Lake sample may play a role here.

The present leads to the development model of settlement along the San Luis Rey River, which, although far from complete, allows some assessment of ancient intrabasin dynamics. La Jolla and Pauma are best understood as functional variants of the same cultural manifestation, reflecting maximal use of a variety of ecological zones to safeguard against environmental hardships. Concentration of groups around the estuarine zone and in the mountainous inland zone is viewed as two poles of one adaptation system. The adaptational balance shifted through time due to environmental constraints. It seems likely that when the coastal lagoon silted up, some 3,500 years ago, La Jolla groups gradually moved inland to exploit a more varied resource zone. It is generally assumed that balanophagous economies are a relatively late adaptation. If further investigations determine that the earliest settlements are located along the coast, then the possibility of lagoon silting causing pressure on local populations and forcing them to focus on other ecozones to relieve the stress, should be investigated.

UNRESOLVED ISSUES

The apparent hiatus between La Jolla/Pauma and San Luis Rey complexes is enigmatic. More dates from a wide variety of sites are needed to assess whether an actual lacuna occurs in the archaeological record. It is likely that this time represents a period of lower populations rather than depopulation of the area. Intrabasin adaptation strategies of La Jolla/Pauma and San Luis Rey groups are strikingly similar. Clearly, the possibility of a continuum between these complexes needs to be further investigated.

For the lower river basin, one of the more urgent archaeological problems is really a geomorphic issue. As long as the extent and character of the ancient estuarine zone remains unknown, our understanding of La Jollan settlement along the river will be blurred. We hope that our work at Whelan Lake has delineated at least some of the steps necessary to help lift the shroud obscuring the panorama of ancient settlement dynamics along the San Luis Rey River.

CHAPTER 12

SUMMARY AND RECOMMENDATIONS

The archaeological testing project at the Whelan Lake site had three major goals: (1) to assess the eligibility of the site for inclusion in the National Register of Historic Places, (2) to determine whether human remains are present, and (3) to develop a treatment plan for the site. This chapter addresses these goals as well as directions for future research.

NATIONAL REGISTER ELIGIBILITY

The eligibility of the Whelan Lake site for inclusion on the National Register of Historic Places (NRHP) hinges on the site's potential to yield data important to the prehistory of the region. To evaluate the site, we formulated questions relating to five research domains: paleoenvironment, chronology, subsistence and settlement, site structure and organization, and exchange and intergroup relations. We argued that if the site could address questions relating to one or more of these topics, then it would be considered eligible for inclusion on the NRHP.

Absolute dates based on six shell assays indicate that the site was occupied between 6,500 and 7,400 B.P. The dates were, for the most part, internally consistent, with the earlier dates deriving from the bottom of the midden and the younger dates obtained from the top. Obsidian hydration measurements, although not easily rendered into absolute dates, were also clustered, suggesting a relatively short-term, single component occupation.

Tenuous support for the absolute dates was provided by the cultural assemblage. The base of a Pinto style projectile point as well as the predominance of *Chione* in the assemblage are consistent with early assemblages in San Diego County. The current leaseholder of the property contends that he has a collection of projectile points from the site. Unfortunately, this collection could not be viewed during the fieldwork, but may be accessible to future researchers.

The early dates for the site gave added importance to issues of subsistence and the environmental context. Both shell and animal bones point to a coastal emphasis in resource procurement. All shell recovered at SDI-6010 is marine in origin. Similarly, the fish remains represent lagoonal and near shore marine species. No freshwater fish remains were found. Aquatic birds, which could inhabit both fresh and salt water marshes, were also found. In general, the vertebrate and invertebrate faunal remains are consistent with a site near the edge of a lagoon or salt water marsh.

The geomorphic evidence, however, does not support such an interpretation. According to Orme (Chapter 2), the sea in the form of a restricted estuary probably did not extend more than 2 km inland during the height of the Flandrian Transgression (ca. 5,000 B.P.), and almost certainly not more than 4 km from the coast. At its maximum extent, then, the ancient lagoon would have been no closer than 1.0 km from SDI-5130 and 3 km from SDI-6010.

The maximum extent of the lagoon, however, would have been around 5,000 B.P., not 6,500 to 7,400 B.P. when the Whelan Lake site was occupied. At 7,000 B.P., sea level would have been between 12 m and 15 m below current levels. The amount of ocean water capable of intruding into the Oceanside Narrows would have been correspondingly less. A lagoon at this time would have been even more restricted in extent than at 5,000 B.P.

Even if a salt water marsh became established above the narrows, it is unlikely that it would be long-lived. About 2 km from the coast, the floodplain of the San Luis Rey River is little more than 150 m wide. With such a narrow constriction connecting a lagoon with the ocean, it is almost assured that any salt water lagoon that could have been established would have quickly become choked with sediment, turning brackish and then fresh. As far upstream as Whelan Lake, it is doubtful that a salt water marsh could have lasted more than few hundred years.

Whelan Lake lies near the juncture of the middle and lower basin. Above SDI-6010, the ridges rise quickly from the river, whereas below the floodplain flattens out as the river makes its way to the sea. Around 7,000 B.P., the floodplain below SDI-6010 probably would have hosted a fresh water lagoon and marsh. This wetland environment would have eventually been filled with alluvial sediment, creating the setting viewed today. At 7,000 B.P., the vicinity around Whelan Lake would have represented the first massive block of dry land as one moved inland from the coast. Indeed, the placement of the mission and the first coastal road by the Spanish reflects this fact.

Perhaps the most perplexing aspect of interpreting the Whelan Lake site is reconciling the geomorphic evidence with the cultural assemblage. Both the artifacts and ecofacts in the collection point to a strong emphasis on coastal resources that according to the geomorphic assessment would have been located between 3 km and 5 km downstream. This coastal orientation is even more intriguing for it appears stronger than the one manifested at SDI-5130, a large village site dating between 6,400 and 4,000 B.P. and located 1.5 km closer to the ocean than SDI-6010.

It is possible that the inhabitants of SDI-6010 were relative newcomers to the coast. They may have migrated from wetland environments in the deserts and Great Basin that had dried up. These environments were probably similar to the fresh water lagoons and marshes encountered around SDI-6010. Perhaps feeling more comfortable closer to potable water, the resources of the marsh, and the relative security of elevated landforms, these people established their camp on the knoll at Whelan Lake, and from there exploited a variety of resources, many of which were located closer to the ocean.

It does not appear that any single occupation of SDI-6010 lasted more than a few months. The relatively restricted diversity of artifacts and lithic technologies present suggest that only part of the seasonal round is represented at the site. Tools flaked from good knappable material appear to have followed a curated strategy in which the preforms were finished when needed and tools were resharpened after their original use life was complete. The lack of cores and finished tools of these materials suggests that only a limited amount of finishing work was conducted at the site, with the tools and cores of these materials being transported to the next site.

Little information was recovered about the time of year the site was occupied. The only birds identified in the faunal assemblage are year-round residents of the area. It is possible that the site was occupied for short periods of time throughout the year rather than for an extended period during one season. Additional research on this topic is warranted.

While the occupation appears to have been seasonal, it also appears to have been recurrent. Based on the vertical distribution of artifacts and ecofacts, the occupation was probably episodic, with periods of relatively intense occupation followed by intervals of sporadic use. The horizontal distribution of lithic raw materials suggests that different areas of the site were utilized for different tool manufacture. Additional support for distinct activity areas is provided by four discrete moist areas in the midden that have been interpreted as processing areas. Two of these areas are associated with hearths that were used for the processing of shell, particularly *Donax*, and probably other resources as well.

The size of the midden indicates that the site was never occupied by more than a small group. These groups appear to have used the site as a base camp from which to exploit resources over a relatively large area. When in residence at the site, occupants apparently travelled about 7 km to the coast for food resources and between 10 km and 20 km to the hills and ridges of the Camp Pendelton area for lithic resources. Interaction with groups further away is indicated by the presence of obsidian from Coso and Obsidian Buttes source areas.

In short, the Whelan Lake site appears to have been used by small groups as a temporary camp either seasonally or periodically throughout the year. Although situated close to freshwater marshes and open lagoons, the inhabitants focused much of their energy on marine and lagoonal resources only available in salt water wetland environments at some distance from the site. The rationale for this settlement paradox is unclear, and, like so much of the cultural outline for this period, remains to be explained.

Data from SDI-6010 have provided valuable information on a poorly understood period of southern California prehistory. The site is the earliest dated component in the San Luis Rey River basin. It retains a high degree of integrity, and can assuredly provide additional important information on the prehistory of the region. For this reason, we recommend that the Whelan Lake site is eligible for inclusion in the NRHP.

HUMAN REMAINS

One fragment of human bone was recovered during the testing of the Whelan Lake site. The piece was identified in the laboratory at which time the COE, San Diego Coroner, the California Native American Heritage Committee, and Henry Rodriguez, the on-site Native American monitor, were contacted. The parties will decide the final disposition of the fragment.

Although only one fragment was found, the testing program was successful in determining that human remains are present at the Whelan Lake site. Given that only a small fraction of the site was excavated, we would anticipate that a considerably larger number of human remains are present at the site.

TREATMENT PLAN

Based on the eligibility of the site for inclusion in the NRHP and the presence of human remains in the deposit, the COE has decided to avoid the prehistoric component during construction. The site has been marked and an archaeologist is monitoring construction to insure that the site will not be impacted by these activities.

DIRECTIONS FOR FUTURE RESEARCH

Although the San Luis Rey River basin remains one of the best studied archaeological areas in southern California, our studies at the Whelan Lake site have brought into sharp focus that there are still major gaps in our knowledge of the region's prehistory. In closing this report we offer what we believe are the next few steps in the cumulative process of reconstructing the area's past.

1. Perhaps the most critical issue to resolve concerns the timing and extent of the ancient lagoon at the mouth of the San Luis Rey River. Orme concludes that the lagoon never extended inland more than 4 km, and that the maximum extent of the lagoon coincided with the Flandrian

Transgression, or about 5,000 B.P. Others believe that the lagoon extended further inland and was in place earlier (Dennis Gallegos, personal communication 1991). Orme's interpretation is based primarily on archival records and maps. He suggests that the problem could be readily resolved by examining the core data from the floodplain which should reveal (1) the location of the bedrock channel, and (2) the location and thickness of estuarine deposits.

2. We hypothesized that the greater quantity and diversity of shellfish remains at the Whelan Lake site as opposed to the larger, downstream village site at SDI-5130 was a reflection of continued use of the mudflats in the lagoon over several thousand years. To test this hypothesis quantitative data on shell size from the two sites should be compared.

3. Virtually no inland La Jollan-Pauma sites have been excavated in the San Luis Rey River basin. Yet, these sites hold the key for understanding the orientation of cultural groups during this period. Competing hypotheses view these manifestations along physiographic lines, with the traditional model delineating a coastal culture and an inland one, as opposed to the model articulated here which suggests that sociocultural entities encompassed entire river basins. Data from inland sites for this period are key to resolving this issue.

4. The Whelan Lake site is the oldest known site in the San Luis Rey River basin. The site retains a high degree of integrity, although several factors are currently impacting it. The site's location on an exposed knoll has left it vulnerable to vandalism. Thus far, vandalism has been restricted to the surface collecting of projectile points and other artifacts. A far more serious threat to the site is the current land use. A sprinkler system was installed throughout the site prior to our project. At the end of the project, the leaseholder indicated his desire to replace the old system with a new one. Such a course would have a major adverse impact on the site.

   The Whelan Lake site represents a rare resource in both its age and integrity of its deposits. The site needs to be protected. Therefore, we recommend that the COE actually nominate the site to the National Register of Historic Places.

REFERENCES

Abbott, R. Tucker
1968 *A Guide to Field Identification Sea Shells of North America*. Western Publishing Company, Inc.

Adams, Karen
1987 Little Barley (*Hordeum pusillum* Nutt.) as a Possible New World Domesticate. In *Specialized Studies in the Economy, Environment and Culture of La Ciudad*, edited by J. Kisselburg, G. Rice and B. Shears, Appendix 9C. Arizona State University Anthropological Field Studies 20(1). Tempe.

Affinis
1991 *Draft Environmental Impact Report for the Whelan Property Acquisition*. Prepared for the City of Oceanside.

Altschul, Jeffrey H.
1991 Archaeological Treatment Plan for CA-SDI-6010. Ms. on file, U.S. Army Corps of Engineers, Los Angeles, California.

Altschul, J.H., M.R. Rose and M.K. Lerch
1984 *Man and Settlement in the Upper Santa Ana River Drainage: A Cultural Resources Overview*. Report prepared for the Los Angeles District, U.S. Army Corps of Engineers, Los Angeles. Statistical Research Technical Series No. 1. Tucson.

Altschul, J.H. and S. Shelley
1987 *Yamisevul: An Archaeological Treatment Plan and Testing Report for CA-RIV-269, Riverside County, California*. Statistical Research Technical Series No. 9. Tucson.

Andermann, G. and J.W. Kemp
1958 Scattered X-rays as Internal Standards in X-ray Emission Spectroscopy. *Analytical Chemistry* 30:1306-1309.

Antevs, E.
1953 On the Division of the Last 20,000 Years. *University of California Archaeological Survey Reports* 22:5-8. Berkeley.

Ariotti, C.
n. d. The History of Whelan Lake Bird Sanctuary.

Asch, David L. and Nancy B. Asch
1985 Prehistoric Plant Cultivation in West-Central Illinois. In *Prehistoric Food Production in North America* edited by Richard I. Ford, pp. 149-203. Anthropological Papers, Museum of Anthropology, University of Michigan No. 75. Ann Arbor.

Axelrod, D.L.
1981 Holocene Climate Changes in Relation to Vegetation Disjunction and Speciation. *American Naturalist* 117(6):847-870.

Bada, J.L., R.A. Schroeder and G.F. Carter
1974 New Evidence for the Antiquity of Man in North America Deduced from Aspartic Acid Racemization. *Science* 184:791-793.

Bancroft, H.H.
1886 *The History of California: The Works of Hubert Howe Bancroft.* The History Company, San Francisco.

Barbour, M.G. and J. Major.
1988 *Terrestrial Vegetation of California.* California Native Plant Society Special Publication 9.

Barker, M., E. Burton and W.M. Childers
1973 A Preliminary Report on a Burial Excavated in the Yuha Desert of Imperial County, California. Ms. on file, Bureau of Land Management, Riverside, California.

Basgall, M.E.
1987 Resource Intensification Among Hunter-Gatherers: Acorn Economies in Prehistoric California. In *Research in Economic Anthropology*, Vol. 9, edited by B.L. Isaac, pp. 21-52.

Basgall, M.E. and D.L. True
1985 *Archaeological Investigations in Crowder Canyon, 1973-1984: Excavations at Sites SBR-421B, SBR-421C, SBR-421D, and SBR-713.* Prepared for California Department of Transportation, District 8, San Bernardino. Far Western Anthropological Research Group, Inc., Davis.

Baumhoff, M.A.
1963 Ecological Determinants of Aboriginal California Populations. *University of California Publications in American Archaeology and Ethnology* 49:155-236. Berkeley.

Bean, L.J. and F.C. Shipek
1978 Luiseno. In *California*, edited by Robert F. Heizer, pp. 538-549. Handbook of North American Indians, vol. 8, William G. Sturtevant, general editor. Smithsonian Institution, Washington, D.C.

Beauchamp, R. Mitchel
1986 *A Flora of San Diego County, California.* Sweetwater River Press, National City, CA.

Beck, W.A. and Y.D. Haase
1974 *Historical Atlas of California.* University of Oklahoma Press, Norman.

Bedwell, S.F.
1970 *Prehistory and Environment of the Pluvial Fort Rock Lake Area, South-Central Oregon.* Unpublished Ph.D. dissertation. Department of Anthropology, University of Oregon.

Behler, John L.
1979 *The Audubon Society Field Guide to North American Reptiles and Amphibians.* Alfred K. Knopf, Inc., New York.

Bettinger, R.L.
1979 Multivariate Statistical Analysis of a Regional Subsistence-Settlement Model for Owens Valley. *American Antiquity* 44(3):455-470.

Bettinger, R.L. (continued)
1980 Explanatory/Predictive Models of Hunter-Gatherer Adaptation. In *Advances in Archaeological Method and Theory* 3, edited by M.B. Schiffer, pp. 189-255. Academic Press, New York.

Bickel, P. McW.
1978 Changing Sea Levels Along the California Coast: Anthropological Implications. *The Journal of California Anthropology* 5(1):6-19.

Binford, Lewis R.
1962 Archaeology as Anthropology. *American Antiquity* 28:217-225.

1968 Archaeological Perspectives. In *New Perspectives in Archaeology.*, edited by S. Binford and L. Binford, pp. 313-341. Aldine Publishing Company, Chicago.

1972 *An Archaeological Perspective.* Seminar Press, New York.

1977 General Introduction. In *For Theory Building in Archaeology*, pp. 1-10. Academic Press, New York.

1978 Dimensional Analysis of Behavior and Site Structure: Learning from an Eskimo Hunting Stand. *American Antiquity* 43:330-361.

1979 Organizations and Formation Processes: Looking at Curated Technologies. *Journal of Anthropological Research* 35:255-272.

1980 Willow Smoke and Dog's Tails: Hunter-Gatherer Settlement Systems and Archaeological Site Formation. *American Antiquity* 45:1-17.

1982 The Archaeology of Place. *Journal of Anthropological Archaeology* 1:5-31.

Bischoff, J.L., W.M. Childers and R.M. Shlemon
1978 Comments on the Pleistocene Age Assignment and Associations of a Human Burial from the Yuha Desert, California: A Rebuttal. *American Antiquity* 43(4):747-749.

Bischoff, J.L., R. Merriam, W.M. Childers and R. Protsch
1976 Antiquity of Man in America Indicated by Radiometric Dates on the Yuha Burial Site. *Nature* 261:128-129.

Bleitz-Sanburg, Dana E.
1987 *The Changing Exploitation of Mammalian and Avian Food Resources at SNI-11, San Nicolas Island, California.* Unpublished Master's thesis in Anthropology, California State University, Los Angeles.

Bock, C.E. and J.H. Bock
1974 Geographical Ecology of the Acorn Woodpecker: Diversity Versus Abundance of Resources. *The American Naturalist*, 108(963):694-698.

Bolton, H.E., editor
1926 *Historical Memoirs of New California.* University of California Press, Berkeley.

Botkin, S.
1980 Effects of Human Exploitation on Shellfish Populations at Malibu Creek, California. In *Modeling Change in Prehistoric Subsistence Economies,* edited by T.K. Earle and A.L. Christenson, pp. 121-139. Academic Press, New York.

Bowers, D.
1976 A Surface Survey of Las Flores Ranch (*Guapiabit*). Los Angeles: *Archaeological Survey Association of Southern California Occasional Papers* 9.

Brooks, Ralph E. and William C. Johnson
1991 Macrobotanical Analysis of the Admiralty Site. In *Life in the Ballona: Archaeological Investigations at the Admiralty Site (CA-LAN-147) and the Channel Gateway Site (CA-LAN-1596-H),* by Jeffrey H. Altschul and Jeffrey A. Homburg, pp. 328-335. Statistical Research Technical Series No. 29. Tucson.

Brownlie, W.R., T.C. Fall and B.D. Taylor
1979 *Sediment Management for Southern California Mountains, Coastal Plains and Shoreline.* California Institute of Technology, Pasadena.

Bull, C.S.
1987 A New Proposal: Some Suggestions for San Diego Prehistory. In *San Dieguito - La Jolla: Chronology and Controversy*, edited by D.R. Gallegos and S.M. Hector, pp. 35-44. San Diego County Archaeological Society Research Paper No. 1. San Diego.

Burt, William H. and Richard P. Grossenheider
1976 *A Field Guide to the Mammals of America North of Mexico.* Houghton-Mifflin, Boston.

Butler, W.B.
1987 Significance and Other Frustrations in the CRM Process. *American Antiquity* 52:820-829.

California Coastal Commission
1987 *California Coastal Resource Guide.* University of California Press, Berkeley.

Campbell, E.W.C., W.H. Campbell, E. Antevs, C.E. Amsden, J.A. Barbieri and F.D. Bode.
1937 The Archaeology of Pleistocene Lake Mojave. *Southwest Museum Papers* No. 11. Los Angeles.

Campbell, I. and G.B. Oakeshott
1966 *Geologic Map of California, Santa Ana Sheet.* State of California, Division of Mines and Geology, Sacramento.

Carbone, L.A.
1991 Early Holocene Environments and Paleoecological Contexts on the Central and Southern California Coast. In *Hunter-Gatherers of Early Holocene Coastal California,* edited by J.M. Erlandson and R.H. Colten, pp. 11-17. Perspectives in California Archaeology, Vol 1. Institute of Archaeology, University of California, Los Angeles.

Carrico, R.L.
1978 Archaeological Excavation and Testing at Site SDi-5130, Oceanside, California. Ms. on file at WESTEC Services, Inc.

Carrico, R.L. (continued)
1987 Sixty-Five Years of San Diego County Archaeology. In *San Dieguito - La Jolla: Chronology and Controversy*, edited D.R. Gallegos and S.M. Hector, pp. 1-14. San Diego County Archaeological Society Research Paper 1. San Diego.

Carter, G.F.
1949 Evidence for Pleistocene Man at La Jolla, California. *New York Academy of Sciences Transactions Series* 2(7):254-257.

1950 Evidence of Pleistocene Man in Southern California. *Geographical Review* 40(1):84-102.

1951 Man in America: A Criticism of Scientific Thought. *The Scientific Monthly* 73(5):297-307.

1952 Interglacial Artifacts from the San Diego Area. *Southwestern Journal of Anthropology* 8(4):444-456.

1954 An Interglacial Site at San Diego California. *Masterkey* 28(5):165-174.

1957 *Pleistocene Man at San Diego.* Johns Hopkins Press, Baltimore.

Cashdan, E.
1990 *Risk and Uncertainty in Tribal and Peasant Economies.* Westview Press, Boulder.

Caughey, J.W.
1953 *California.* Prentice-Hall, New York.

Caughman, M.M. and J.S. Ginsberg, editors
1981 *California Coastal Access Guide.* California Coastal Commission, University of California Press, Berkeley.

Cerreto, Richard and Michael A. Foertsch
1985 An Illustrated Identification Key for Shellfish Recovered from Archaeological Sites I. Orange County. In *Pacific Coast Archaeological Society Quarterly*, Vol. 21, No. 3, July 1985.

Chartkoff, Joseph L. and Kerry Kona Chartkoff
1984 *The Archaeology of California.* Stanford University Press, Stanford.

Childers, W.M.
1974 Preliminary Report on the Yuha Burial, California. *Anthropological Journal of Canada* 12(1):2-9.

Childers, W.M. and H.L. Minshall
1980 Evidence of Early Man Exposed at Yuha Pinto Wash. *American Antiquity* 45(2):297-308.

Colson, E.
1979 In Good Years and in Bad: Food Strategies of Self-Reliant Societies. *Journal of Anthropological Research* 35(1):18-29.

Conkey, M. and J. Spector
1984 Archaeology and the Study of Gender. In *Advances in Archaeological Method and Theory*, vol. 7, edited by M. Schiffer. Academic Press, New York.

Cook, S.F.
1976 *The Population of the California Indians 1769-1970.* University of California Press, Berkeley.

Corps of Engineers
1991 Scope of Work, Contract DACW09-91-D-0009, Statistical Research Inc., Delivery Order No. 1. Los Angeles District, Corps of Engineers, Los Angeles.

Crabtree, Don E.
1972 *An Introduction to Flint Working.* Occasional Papers of The Idaho State Museum No. 29. Pocatello.

Crabtree, R.H., C.N. Warren and D.L. True
1963 Archaeological Investigations at Batiquitos Lagoon, San Diego County, California. Report prepared for the California Division of Beaches and Parks and the Division of Highways.

Crouch, H.W.
1915 Recollections and Bibliographical Notes. San Diego Historical Society Research Archives.

Cultural Resource Management Services
1989 *Draft Field Plan: Test Excavations at San Luis Rey.* Prepared for Los Angeles District, Corps of Engineers, Los Angeles, California.

Cupples, Sue Ann
1976 *Oceanside Harbor and Navigation Project: Archaeological Survey Report.* Prepared for U.S. Army Corps of Engineers. San Diego State University Foundation, San Diego, California.

Cupples, Sue Ann and Ken Hedges
1977 *San Luis Rey River Basin: Overview of Cultural Resources.* Report prepared for U.S. Army Corps of Engineers. San Diego State University Foundation, San Diego, California.

Dalton, G.
1961 Economic Theory and Primitive Society. *American Anthropologist* 63:1-25.

Davis, O.K.
1989 Section I - Pollen analysis of San Joaquin Marsh. Report to Chambers Group, Santa Anna California. 25 pp.

1990 Preliminary report of the pollen analysis of Tulare Lake. TULARG 3(5):2-4.

1991 Pollen analysis of CA-LAN-60, Los Angeles, California. Report submitted to Jeffrey A. Homburg, Statistical Research, Inc., Tucson, Arizona. 10 pp.

Davis, O.K. and C.H. Miksicek
1988a Pollen, Charred Seeds, and Wood Charcoal from FRE-64. Report submitted to Chambers Group, Santa Anna California. 20 pp.

1988b Pollen, Charred Seeds, and Wood Charcoal from the Redbank-Fancher Creeks Project. Report submitted to Chambers Group, Santa Anna California. 31 pp.

1988c Archeobotanical Analyses. In *Archaeological Excavations at Site CA-FRE-1671, Fresno County, California: Final Report*, by Michael J. Moratto, pp. N.1-N.6. Submitted to California Department of Transportation.

Deevy, E.S. and R.F. Flint
1957 Postglacial Hypsithermal Interval. *Science* 125(3240):182-184.

Desgrandchamp, C.S.
1976 *Archaeological Investigations at Elkhorn Slough: Ca-MNT-229, A Middle Period Site on the Central California Coast.* Papers in Northern California Anthropology 3. Northern California Anthropological Group, Berkeley.

Dixon, K.A.
1968 Cogged Stones and Other Ceremonial Cache Artifacts in Stratigraphic Context at ORa-58, a Site in the Lower Santa Ana River Drainage, Orange County. *Pacific Coast Archaeological Society Quarterly* 4(3):57-65.

1975 New Evidence for the Most Important Archaeological Discovery in Long Beach: The Cogged Stones and Discs of Rancho Los Cerritos. *Los Fierros* 12(2):20-31.

Donnan, C.B.
1964 A Suggested Culture Sequence for the Providence Mountains (Eastern Mojave Desert). In *Archaeological Survey Annual Report for 1963-1964*, Vol. 6, pp. 1-27. University of California, Los Angeles.

Driver, H.E.
1953 The Acorn in North American Indian Diet. *Proceedings of the Indiana Academy of Sciences* 62:56-62.

Drover, C.E.
1977 Archaeological Site Forms on File with San Diego State University and the San Diego Museum of Man.

Du Bois, C.G.
1904 Mythology of the Mission Indians. *Journal of American Folk-Lore* 17(66):185-188.

1908 The Religion of the Luiseno and Diegueno Indians of Southern California. *University of California Publications in American Archaeology and Ethnology* 8(3):69-186. Berkeley.

Dunbar, C.O. and K.M. Waage
1969 *Historical Geology* (Third Edition). John Wiley and Sons, Inc., New York.

Dunnell, R.C.
1982 Science, Social Science and Common Sense: The Agonizing Dilemma of Modern Archaeology. *Journal of Anthropological Research* 38:1-25.

Dyer, K.R.
1973 *Estuaries: A Physical Introduction*. New York: John Wiley and Sons.

Ebeling, Walter
1986 *Handbook of Indian Foods and Fibers of Arid America.* University of California Press, Berkeley.

Eberhart, H.
1952 Report on an Archaeological Survey near Rincon Indian Reservation, San Diego County, California. Ms. on file, Institute of Anthropology, University of California, Los Angeles.

Eberhart, H. (continued)
1961 The Cogged Stone of Southern California. *American Antiquity* 26(3):361-370.

Eberhart, H. and W. Wasson
1975 The Sasson Site (LAn-339), a Milling Stone Horizon Station in the San Gabriel Valley, California. *California Anthropologist* 5:9-45.

Elliot, W.W.
1883 *History of San Bernardino and San Diego Counties, California.* Riverside Museum Press, 1967.

Emery, K.
1967 Estuaries and Lagoons in Relation to Continental Shelves. In *Estuaries*, edited by G.H. Lauff, 9-14. American Association for the Advancement of Science 83. Washington, D.C.

Engelhardt, Z.
1921 *San Luis Rey Mission*. The James H. Barry Company, San Francisco.

Engstrand, I. and M. Ward
n.d. Rancho Guajome, An Architectural History. Ms. on file, County of San Diego, Department of Parks and Recreation, San Diego.

Erlandson, J.M.
1991 Shellfish and Seeds as Optimal Resources: Early Holocene Subsistence on the Santa Barbara Coast. In *Hunter-Gatherers of Early Holocene Coastal California,* edited by J.M. Erlandson and R.H. Colten, pp. 89-111. Perspectives in California Archaeology, Vol. 1. Institute of Archaeology, University of California, Los Angeles, .

Erlandson, Jon McVey
1988 *Of Millingstones and Molluscs: The Cultural Ecology of Early Holocene Hunter-Gatherers on the California Coast*. Unpublished Ph.D. dissertation, University of California Santa Barbara.

Erlandson, J.M. and R.H. Colten
1991 An Archaeological Context for Early Holocene Studies on the California Coast. In *Hunter-Gatherers of Early Holocene Coastal California*, edited by J.M. Erlandson and R.H. Colten, pp.1-10. Perspectives in California Archaeology, Vol 1. Institute of Archaeology, University of California, Los Angeles.

Erlandson, J.M. and R.H. Colten, editors
1991 *Hunter-Gatherers of Early Holocene Coastal California.* Perspectives in California Archaeology, Vol 1. Institute of Archaeology, University of California, Los Angeles.

Ezell, P.H.
1977 *An Archaeological Survey of The Santa Margarita River Valley and Adjacent Areas, Camp Pendleton, San Diego County, California.* Department of Anthropology, San Diego State University.

1987 The Harris Site - An Atypical San Dieguito Site or am I Beating a Dead Horse? In *San Dieguito - La Jolla: Chronology and Controversy,* edited by D.R. Gallegos and S.M. Hector, pp. 15-22. San Diego County Archaeological Society Research Paper No. 1.

Fay, J.S., A.G. Lipow and S.W. Fay, editors
1987 *California Almanac* (Third Edition). Pacific Data Resources, Santa Barbara.

Fink, Gary R.
1974 *The Archaeological Resources of Guajome Regional Park. Oceanside, California (565 Acres).* County of San Diego, Public Works Agency, County Engineer Department, San Diego.

1980 *Rancho Guajome: Window on the Past, a Test of the Historic Resources at the Casa de Rancho Guajome.* County of San Diego, County Transportation Department, San Diego.

Fladmark, K.R.
1979 Routes: Alternate Migration Corridors for Early Man in North America. *American Antiquity* 44:55-69.

Fowler, P.
1977 *Approaches to Archaeology*. Adam and Charles Black, London.

Franklin, R.L. and R.L. Carrico
1978 *A Preliminary Archaeological Reconnaissance for a Proposed Flood Control Project in the Lower San Luis Rey River Drainage.* Submitted to the Los Angeles District, Corps of Engineers, Los Angeles, California.

1979 *Cultural Resource Test Sampling Program for a Proposed Flood Control Project in the Lower San Luis Rey River Drainage, Oceanside California.* Submitted to the Los Angeles District, Corps of Engineers, Los Angeles, California.

Freeman, T.A.
1991 Chronometric Determinations for the Northern Del Rey Hills, Los Angeles County, California. *Pacific Coast Archaeology Society Quarterly* 27(1):1-11.

Freeman, T.A. and D.M. Van Horn
1990 Salvage Excavations at the Walker Ranch: A Portion of a Late Prehistoric and Historic Luiseno Village (CA-RIV-333). *Pacific Coast Archaeological Society Quarterly* 26:1-50.

Gallegos, D.R.
1985 Batiquitos Lagoon Revisited. *Casual Papers* 2(2):1-13. Cultural Resource Management Center, San Diego State University, San Diego.

1987 A Review and Synthesis of Environmental and Cultural Material for the Batiquitos Lagoon Region. In *San Dieguito - La Jolla: Chronology and Controversy.* Pp. 23-34. San Diego County Archaeological Society Research Paper No. 1.

1991 Antiquity and Adaptation at Agua Hedionda, Carlsbad, California. In *Hunter-Gatherers of Early Holocene Coastal California*, edited by J.M. Erlandson and R.H. Colten, pp.19-41. Perspectives in Archaeology, University of California, Los Angeles.

Gallegos, D.R. and R. Carrico
1984 *Windsong Shores Data Recovery Program for Site W-131 Carlsbad, California Report Prepared for Hunts Partnership, Carlsbad, California.* Report prepared by WESTEC Services, Inc., San Diego, California.

Gallegos, D.R. and S.M. Hector, editors
1987 *San Dieguito - La Jolla: Chronology and Controversy.* San Diego County Archaeological Society Research Paper No. 1. San Diego.

Gibson, E.C., M.J. Moratto, J.M. Foster, R.S. Greenwood, D.A. Sandler, G.R. Romani, and C.B. Hemphill.
1990 *Archaeological Investigations at Five Sites on the Lower San Luis Rey River, San Diego County, California.* Prepared for Los Angeles District, Corps of Engineers, INFOTEC Research, Inc., Fresno.

Gibson, Eric C. and Debra A. Sandler
1990 Appendix E: Preliminary Description and Analysis of Cultural Materials From Archaeological Sites on The Lower San Luis Rey River, San Diego County, California. In *Archaeological Investigations at Five Sites on The Lower San Luis Rey River, San Diego County, California,* by Eric C. Gibson, M.J. Moratto, J.M. Foster, R.S. Greenwood, D.A. Sandler, G.R. Romani, and C.B. Hemphill, pp. E.1-E.53. Prepared for Los Angeles District Corps of Engineers. INFOTEC Research, Inc. Fresno.

Gifford, E.W.
1918 Clans and Moieties in Southern California. *University of California Publications in American Archaeology and Ethnology* 14(2):155-219. Berkeley.

1922 California Kinship Terminologies. *University of California Publications in American Archaeology and Ethnology* 18(1):1-285. Berkeley.

1936 Californian Balanophagy. In *Essays Presented to A.L. Kroeber*, edited by R. Lowie, pp. 87-98. University of California Press, Berkeley.

Glassow, M.A., L. Wilcoxon and J.M. Erlandson
1988 Cultural and Environmental Change During the Early Period of Santa Barbara Channel Prehistory. In *The Archaeology of Prehistoric Coastlines*, edited by Bailey and Parkington, pp. 64-77. Cambridge University Press, Cambridge.

Goodyear, A.C.
1979 *A Hypothesis For The Use of Cryptocrystalline Raw Materials Among The Paleo-Indian Groups of North America.* Research Manuscript No. 156. University of South Carolina, Columbia.

Gotshall, Daniel W.
1989 *Pacific Coast Inshore Fishes*, edited by Kenneth Hashagen. Sea Challenges, Monterey, California.

Grayson, Donald K.
1984 *Quantitive Zooarchaeology: Topics in the Analysis of Archaeological Faunas*. Academic Press, New York.

Greenwood and Associates
1991 Report on Data Recovery at CA-SDI-5130/5133H/6015 and CA-SDI-6014, Historical Components. Greenwood and Associates/INFOTEC Research. Report in preparation.

Greenwood, R.S.
1978 *A Report of Heritage Resources Survey, Phase II, for a Proposed Highway Widening Project on Route 76 in Oceanside, 11-SD-76, P.M. 3.6-4.2.* Submitted to the California Department of Transportation, San Diego.

Griner, E.L. and P.R. Pryde
1976 Climate, Soils, and Vegetation. In *San Diego: An Introduction to the Region*, edited by P. R. Pryde, pp. 29-46. Kendall/Hunt Publishing Co, Dubuque.

Hammett, Julia E.

1989a *Analysis of Carbonized Plant Remains from Site SBA-1816.* Technical report submitted to Martin Marietta Corporation by Environmental Solutions, Inc.

1989b *Analysis of Plant Remains from CA-SLO-165.* Technical report submitted to Director of Public Works, City of Morro Bay, San Luis Obispo County, CA by Singer and Associates.

1990 Analysis of Plant Remains from the 1987 Excavations at Talepop (CA-LAN-229). In *Archaeological Studies at Site CA-LAN-229, Malibu Creek State Park: An Experiment in Inference Justification*, edited by L. Mark Raab, Appendix 2. Northridge Center for Public Archaeology, California State University at Northridge.

Hanes, Ted L.

1976 Vegetation Types of the San Gabriel Mountains. In *Plant Communities of Southern California*, edited by June Latting, pp. 65-76. California Native Plant Society Special Publication No. 2. Los Angeles.

Harding, M.

1951 La Jollan Culture. *El Museo* 1(1):10-11; 31-38. San Diego.

Harrington, J.P.

1933 Fieldwork among the Mission Indians of California. In *Explorations and Fieldwork of the Smithsonian Institution in 1932*, pp. 85-88. Washington.

1934 Rescuing the Early History of the California Indians. In *Explorations and Fieldwork of the Smithsonian Institution in 1934*, pp. 54-56. Washington.

Harrington, Mark Raymond

1957 *A Pinto Site at Little Lake, California.* Southwest Museum Papers No. 17. Los Angeles.

Harry, Karen G.

1990 Chapter V, Lithic Analysis. In *Prehistoric Adaptation To a Desert Spring Environment: Archaeological Investigations of Surprise Spring, San Bernardino County, California*, by Jeffrey A. Altschul, pp. 53-83 and 123-127. Statistical Research Technical Series No. 27. Tucson.

Hayden, J.D.

1966 Restoration of the San Dieguito Type Site to Its Proper Place in the San Dieguito Sequence. *American Antiquity* 31(3):439-440.

1987 Notes on the Apparent Course of San Dieguito Development. In *San Dieguito - La Jolla: Chronology and Controversy*, edited by D.R. Gallegos and S.M. Hector, pp. 45-50. San Diego County Archaeological Society Research Paper No. 1.

Haynes, C.V., Jr.

1969 A Scientist Disagrees... *Encyclopaedia Britannica Yearbook of Science and the Future*, 1970:76-77. Chicago

1973 The Calico Site: Artifacts or Geofacts? *Science* 181(4097):305-310.

Hector, S.M. and W.R. Manley

1987 *Archaeological Test Excavations at Rancho Guajome, County of San Diego, California.* Submitted to County of San Diego, Department of Parks and Recreation, San Diego.

Hector, S.M. and W.R. Manley (continued)
1988 *Results of a Monitoring Program Conducted for Trenching at Rancho Guajome, County of San Diego.* Submitted to County of San Diego, Department of Parks and Recreation, San Diego.

Heizer, Robert F. and Albert B. Elsasser
1980 *The Natural World of the California Indians*. University of California Press, Berkeley.

Henshaw, H.W.
1972 The Luiseno Creation Myth. R.F. Heizer, editor. *Masterkey* 46(3):93-100.

Herskovits, M.J.
1940 *The Economic Life of Primitive Peoples*. Knopf, New York.

Hodder, I.
1985 Post Processual Archaeology. In *Advances in Archaeological Method and Theory*, vol. 2, edited by M. Schiffer. Academic Press, New York.

1986 *Reading the Past*. Cambridge University Press, Cambridge.

Hubbert, B.F.
1936 Memoirs, Biographical File, San Diego Historical Society Research Archives.

Hughes, R.E. and D.L. True
1985 Perspectives on the Distribution of Obsidians in San Diego County, California. *North American Archaeologist* 6(4):325-339.

Inman, D.
1983 Application of Coastal Dynamics to the Reconstruction of Paleocoastlines in the Vicinity of La Jolla, California. In *Quaternary Coastlines and Marine Archaeology*, edited by P.M. Masters and N.C. Flemming, pp. 1-49. Academic Press, New York.

Jackson, B.J.
1977 *Plane Sense: A Technological and Functional Analysis of a Stone Tool Category.* Unpublished Master's thesis, Department of Anthropology, Washington State University, Pullman.

Jehl, J.
1989 Trophic Overview of Whelan Lake. Prepared for Affinis. In *Draft Environmental Impact Report for the Whelan Property Aquisition*. 1991.

Jochim, M.A.
1976 *Hunter-Gatherer Subsistence and Settlement: A Predictive Model.* Academic Press, New York.

Johnson, K.L.
1966 *Site LAn-2, a Late Manifestation of the Topanga Complex in Southern California Prehistory.* University of California Anthropological Records No. 23. Berkeley.

Jones, N.V. and W.J. Wolff
1980 *Feeding and Survival Estuarine Organisms.* Plenum Press, New York.

Jones, T.L.
1991 Marine-Resource Value and the Priority of Coastal Settlement: A California Perspective. *American Antiquity* 56(3):419-443.

Kaldenberg, R.L.
1976 *Paleo-Technological Change at Rancho Park North, San Diego County, California.* Unpublished Master's thesis, Department of Anthropology, California State University, Sacramento.

1982 *Rancho Park North: A San Dieguito-La Jolla Shellfish Processing Site in Coastal Southern California.* Imperial Valley College Museum Society Occasional Papers 6. El Centro, California.

Kaldenberg, R.L. and P.H. Ezell
1974 Results of the Archaeological Mitigation of Great Western Sites A and C, Located on the Proposed Rancho Park North Development near Olivenhain, California. Ms. on file with San Diego State University, Department of Anthropology.

Kelley, J.H. and M.P. Hanen
1988 *Archaeology and the Methodology of Science.* University of New Mexico Press, Albuquerque.

Kennish, M.J.
1986 *Ecology of Estuaries. Volume I: Physical and Chemical Aspects.* CRC Press, Boca Raton.

Ketchum, B.H.
1983 *Ecosystems of the World, 26: Estuaries and Enclosed Seas.* Elsevier Scientific Publishing Company, New York.

King, C.D.
1981 *The Evolution of Chumash Society.* Ph.D. dissertation, Department of Anthropology, University of California, Davis. Ann Arbor: University Microfilms.

Knox, G.A.
1986 *Estuarine Ecosystems: A Systems Approach.* Volumes I and II. CRC Press, Boca Raton.

Koerper, H.C.
1981 *Prehistoric Subsistence and Settlement in the Newport Bay Area and Environs, Orange County, California.* Unpublished Ph.D. dissertation, Department of Anthropology, University of California, Riverside.

Koerper, H.C., J.S. Killingley and R.E. Taylor
1986 The Little Ice Age and Coastal Southern California Human Economy. *Journal of California and Great Basin Anthropology* 7(1):99-103.

Koerper, H.C., P.E. Langenwalter and A. Schroth
1991 Early Holocene Adaptations and the Transitional Phase Problem: Evidence from the Allan O. Kelly Site, Agua Hedionda Lagoon. In *Hunter-Gatherers of Early Holocene Coastal California,* edited by J.M. Erlandson and R.H. Colten, pp. 43-62. Perspectives in California Archaeology, Vol 1. Institute of Archaeology, University of California, Los Angeles.

Kowta, M.
1969 *The Sayles Complex: A Late Milling Stone Assemblage from Cajon Pass and the Ecological Implications of Its Scraper Planes.* University of California Publications in Anthropology No. 6. Berkeley.

Kroeber, A.L.
1906 Two Myths of the Mission Indians of California. *Journal of American Folk-Lore* 19(75):309-321.

1908 Notes on the Luiseno. In "The Religion of the Luiseno Indians of Southern California," by C.G. Du Bois, pp. 174-186. *University of California Publications in American Archaeology and Ethnology* 6(3):369-380. Berkeley.

1909 Classificatory Systems of Relationship. *Journal of the Royal Anthropological Institute of Great Britain and Ireland* 39:77-84. London.

1917 California Kinship Systems. *University of California Publications in American Archaeology and Ethnology* 12(9):336-396. Berkeley.

1925 *Handbook of the Indians of California*. Bureau of American Ethnology. Dover Publications, Inc. 1976, New York.

Kyle, C., R.L. Phillips and D.R. Gallegos
1988 *Cultural Resource Testing for Sites SDi-5130, SDi-6010, and SDi-6010-H for the Oceanside Annexation Project, Oceanside, California*. Prepared by WESTEC Services, Inc. San Diego, California.

Leakey, L.S.B., R.D. Simpson and T. Clements
1968 Archaeological Excavations in the Calico Mountains, California: Preliminary Report. *Science* 160:1022-1023.

Leakey, L.S.B., R.D. Simpson, T. Clements, R. Berger, J. Witthoft and participants of the Calico Conference
1972 *Pleistocene Man at Calico: A Report on the International Conference on the Calico Mountains Excavations, San Bernardino County, California*. San Bernardino County Museum, Redlands.

LeClair, E.E., Jr. and H.K. Schneider
1968 *Economic Anthropology*. Holt, Rinehart and Winston, Inc, New York.

Leone, M.
1981 Archaeology's Relationship to the Present and the Past. In *Modern Material Culture: The Archaeology of Us*, edited by R. Gould and M. Schiffer. Academic Press, New York.

Lynch, H.B.
1931 *Rainfall and Stream Run-off in Southern California Since 1769*. Metropolitan Water District of Southern California, Los Angeles.

Magalousis, N. and H. Kelsey
1990 *Preliminary Archaeological and Historical Investigations at Mission San Luis Rey, California, CA-SDI-241, Sectors A & B*. 3 Vols. Submitted to Parish and Mission of San Luis Rey, Oceanside.

Martin, Alexander C. and William D. Barkley
1973 *Seed Identification Manual*. University of California Press, Berkeley.

Martin, P.S.
1971 The Revolution in Archaeology. *American Antiquity* 36:1-8.

Mead, George R.
1972 *The Ethnobotany of the California Indians*. Museum of Anthropology, University of Northern Colorado, Occasional Publications in Anthropology, Ethnology Series No. 30. Greeley, Colorado.

Meighan, C.W.
1954 A Late Complex in Southern California Prehistory. *Southwestern Journal of Anthropology* 10(2):215-227.

1959a Archaeological Resources of Borrego State Park. *University of California Archaeological Survey, Annual Report 1958-1959*. Los Angeles.

1959b California Cultures and the Concept of an Archaic Stage. *American Antiquity* 24(3): 289-305.

1959c The Little Harbor Site, Catalina Island: An Example of Ecological Interpretation in Archaeology. *American Antiquity* 24(4):383-405.

1978 Obsidian Dating of the Malibu Site. In *Obsidian Dates II: A Compendium of the Obsidian Hydration Determinations Made at the UCLA Obsidian Hydration Lab.* University of California Institute of Archaeology Monograph VI:158-161. Los Angeles.

1989 The Earliest Shell-Mound Dwellers of Southern California. Paper presented at the Circum-Pacific Prehistory Conference, Seattle.

McIntire Group
1990 HEC 2 Modeling, Pilgrim Creek-Improved Channel Report, prepared for the City of Oceanside by the McIntire Group, July 1990.

McLean, James H.
1978 *Marine Shells of Southern California*. Natural History Museum of Los Angeles County, Science Series 24, Revised Edition: 1-104.

McMinn, Howard E.
1974 *An Illustrated Manual of California Shrubs*. University of California Press, Berkeley.

Miksicek, Charles H.
1990 Charred plant remains from sites in the San Joaquin Hills Transportation Corridor. In *Final Test Investigation Report and Request for Determination of Eligibility for 23 Sites along the San Joaquin Hills Transportation Corridors*, edited by Philip de Barros and Henry C. Koerper, Appendix H. Chambers Group, Inc., Santa Ana.

1991 Beyond Acorns, Anadromus Fish, and Artiodactyls: The Potential for Archaeobotany in California Archaeology. Paper presented at the 56th Annual Meeting of the Society for American Archaeology, New Orleans, LA.

Miles, Anne
1978 *Photomicrographs of World Woods*. Her Majesty's Stationery Office, London.

Miller, J.
1966 *The Present and Past Molluscan Faunas and Environments of Four Southern California Coastal Lagoons.* Unpublished Master's thesis, University of California, San Diego.

Moore, J.D.
1985 Shells, Seeds, and Sites. Paper presented at the annual meeting of the Society for California Archaeology, San Diego.

Moratto, M.J.
1984 *California Archaeology*. Academic Press, New York.

Moratto, M.J. and R.S. Greenwood
1991 *Historic Properties Treatment Plan San Luis Rey River Flood-Control Project, San Diego County, California*. Prepared for Los Angeles District, Corps of Engineers, Los Angeles, California.

Moriarty, J.R., III
1966 Cultural Phase Divisions Suggested by Typological Change Coordinated with Stratigraphically Controlled Radiocarbon Dating in San Diego. *The Anthropological Journal of Canada* 4(4):20-30.

1967 Transitional Pre-Desert Phase in San Diego County, California. *Science* 155(3762):553-556.

1969 The San Dieguito Complex: Suggested Environmental and Cultural Relationship. *Anthropological Journal of Canada*, 7(3):1-18.

1987 A Separate Origins Theory for Two Early Man Cultures in California. In *San Dieguito - La Jolla: Chronology and Controversy*, edited by D.R. Gallegos and S.M. Hector, pp. 51-62. San Diego County Archaeological Society Research Paper No. 1. San Diego.

Moriarty, J.R. and R. Broms
1971 The Antiquity and Inferred Use of Stone Discoidals in the Southwest. *The Anthropological Journal of Canada* 10(3):10-13.

Moseley, M. and G.A. Smith
1962 Archaeological Investigations of the Mojave River Drainage. *San Bernardino County Museum Association Quarterly* 9:3.

Moyer, C.C.
1969 *Historic Ranchos of San Diego*. Union-Tribune Publishing.

Mudie, P.J.
1976 Ecological and Taxonomic Diversity in the Modern Saltmarsh Floras of Southern California and Western Baja California. Paper Presented at the Second Southern California Botanist's Symposium. California Native Plant Society.

Munz, Philip A.
1963 *A Flora of Southern California*. University of California Press, Berkeley.

Musil, Albina F.
1963 *Identification of Crop and Weed Seeds*. U.S. Department of Agriculture Handbook No. 219. Washington, D.C.

National Oceanic and Atmospheric Administration
1990 *Climatological Data Annual Summary, California 1990*. National Climatic Data Center, Asheville.

National Park Service
1986 *Guidelines for Completing National Register of Historic Places Forms.* Bulletin No. 16. Register of Historic Places. Department of the Interior, Washington, D.C.

Nielsen, K.K.
1979 Progress in X-ray Fluorescence Correction Methods Using Scattered Radiation. In *Advances in X-ray Analysis*, edited by G.J. McCarthy, vol. 22, pp. 303-315. Plenum Press, New York.

Noah, A.C.
1986 *1986 Archaeological Test Excavations at Rancho Guajome.* County of San Diego, Public Works Agency, County Engineer Department, San Diego.

Norwood, R. and C. Walker
1980 The Cultural Resources of San Dieguito Estates. Report Prepared by and on file with RECON, San Diego, California.

Orme, A.R.
1990 Wetland Morphology, Hydrodynamics and Sedimentation. In *Wetlands: A Threatened Resource*, edited by M. Williams, pp. 42-94. Blackwell, Oxford.

Polanyi, K.
1947 Our Obsolete Market Mentality. *Commentary* 13:109-117.

Pourade, R.F., editor
1966 *Ancient Hunters of the Far West*. Union-Tribune, San Diego.

Quillen, D.K., R.L. Carrico and D. Gallegos
1984 *Archaeological Investigations at SDi-5130, Mar Lado Project, Oceanside, California.* Prepared by WESTEC Services, Inc.

Rathje, W.L. and M.B. Schiffer
1982 *Archaeology*. Harcourt Brace Jovanovich, New York.

RECON
1977 Whelan Ranch Archaeological Survey. Ms. on file with the City of Oceanside.

Rehder, Harold A.
1981 *The Audubon Society Field Guide to North American Seashells*. Alfred A. Knopf, New York.

Reid, J.J.
1991 Editor's Corner: On the Language of Archaeology. *American Antiquity* 56(3):387-388.

Ricketts, Edward F., Jack Calvin and Joel W. Hedgpeth
1985 *Between Pacific Tides*. Fifth edition, revised by David W. Phillips. Stanford University Press, Stanford, California.

Robinson, Stephen W. and Gail Thompson
1981 Radiocarbon Corrections for Marine Shell Dates with Application to Southern Pacific Northwest Coast Prehistory. *Syesis* 14:45-57. The British Columbia Provincial Museum.

Rogers, M.J.
1929a Archaeological Field Work in North America During 1928, California. *American Anthropologist* 31(3):341.

Rogers, M.J. (continued)
1929b *Report on an Archaeological Reconnaissance in the Mojave Sink Region*. San Diego Museum of Man Papers No. 1. San Diego.

1929c The Stone Art of the San Dieguito Plateau. *American Anthropologist* 31(3):454-467.

1938 Archaeological and Geological Investigations in an Old Channel of the San Dieguito Valley. *Carnegie Institution of Washington, Yearbook* 37:344-345. Washington, D.C.

1939 *Early Lithic Industries of the Lower Basin of the Colorado River and Adjacent Desert Areas*. San Diego Museum of Man Papers No. 3. San Diego.

1945 An Outline of Yuman Prehistory. *Southwestern Journal of Anthropology* 1(2):167-198. Albuquerque.

1966 *Ancient Hunters of the Far West*. The Union-Tribune Publishing Company, San Diego, California.

Ross, L.A.
1969 *The Irvine Complex: A Late Prehistoric Horizon Archaeological Complex for the Newport Bay Area, California*. Unpublished Master's thesis, Department of Anthropology, Washington State University, Pullman.

1970 ORa-190: A Descriptive Report of a Late Prehistoric Horizon Site in Orange County, California. *Pacific Coast Archaeological Society Quarterly* 6(2-3):1-135.

Rowlands, M.
1984 Objectivity and Subjectivity in Archaeology. In *Marxist Perspectives in Archaeology*, edited by M. Spriggs. Cambridge University Press, Cambridge.

Sahlins, M.D.
1974 *Stone Age Economics*. London: Tavistock Publications.

Salls, Roy
1983 *The Liberty Grove Site: Archaeological Interpretations of a Late Millingstone Site on the Cucamonga Plain.* Unpublished Master's thesis, Department of Anthropology, California State University, Los Angeles.

1988 *Prehistoric Fisheries of the California Bight.* Unpublished Ph.D. dissertation, Archaeology Interdepartmental Program, University of California, Los Angeles.

Salls, Roy, Dana E. Bleitz-Sanburg and Mercedes Duque
1988 *Prehistoric Faunal Exploitation at the Gretna Green Site (CA-SDI-5426) San Diego County, California*. Anthropological paper of the University of California Institute of Archaeology. Submitted to Archaeological Associates Ltd., Sun City, California.

Sandefur, Elsie C., Susan Colby, Dana Bleitz, Mimi Horner, Lady Harrington, and E. Stepans
1990 Analysis of Faunal Remains from Archaeological Sites in the San Luis Rey River Drainage, San Diego County, California. In *Archaeological Investigations at Five Sites on the Lower San Luis Rey River, San Diego County, California*, by E.C. Gibson, M.J. Moratto, J.M. Foster, R.S. Greenwood, D.A. Sandler, G.R. Romani and C.B. Hemphill. Prepared for Los Angeles District, Corps of Engineers, INFOTEC Research, Inc., Fresno.

San Diego County
1854 Tax Assessment List. On file, San Diego County Historical Society, San Diego.

1867 Tax Assessment List. On file, San Diego County Historical Society, San Diego.

San Luis Rey Historical Society
1957 Grave of Francisco de Ulloa and Prehsitoric Indian Petroglyphs at Oceanside, San Diego County, California. *Bulletin* 9.57. San Luis Rey Historical Society, Oceanside.

Schiffer, M.B.
1976 *Behavioral Archaeology*. Academic Press, New York.

1983 Toward the Identification of Formation Processes. *American Antiquity* 48:675-706.

1987 *Formation Processes of the Archaeological Record*. University of New Mexico Press, Albuquerque.

Schmidt, James J., Mark T. Swanson, John M. Foster and Roberta S. Greenwood
1992 *Historical Archaeological Investigations at CA-SDI-6010/H, San Diego County*. Prepared for U.S. Army Corps of Engineers, Los Angeles District. Greenwood and Associates, Pacific Palisades.

Schreiber, R.W., editor
1981 The Biota of the Ballona Region, Los Angeles County. Supplement 1, Marina Del Rey/ Ballona Local Coastal Plan, Los Angeles County Natural History, Museum Foundation.

Schuiling, W.C., editor
1979 *Pleistocene Man At Calico* (2nd edition). San Bernardino County Museum Association, Redlands.

Shanks, M. and C. Tilley
1987 *Social Theory and Archaeology*. University of New Mexico Press, Albuquerque.

Sharp, R.P.
1972 *Geology Field Guide to Southern California*. Kendall/Hunt, Dubuque.

Shumway, G., C.L. Hubbs and J.R. Moriarty
1961 Scripps Estate Site, San Diego, California: A La Jolla Site Dated 5460 to 7370 Years Before the Present. *Annals of the New York Academy of Science* 93(3):37-132.

Simpson, R.D.
1980 The Calico Mountains Site (Oldest Known Early Man Site in America). *ASA Journal* 4(2):8-25. Archaeological Survey Association of Southern California, Redlands.

Singer, C.A.
1977 Classification and Isolation of the Calico Lithic Industries. In *Abstracts of Papers Presented at the Annual Meeting of the American Ethnological Society for California Archaeology, and Other Organizations* pp. 109. San Diego.

Smith, B.F.
1987 A Reinterpretation of the Transitional Phase. In *San Dieguito - La Jolla: Chronology and Controversy*, edited by D.R. Gallegos and S.M. Hector, pp. 63-74. San Diego County Archaeological Society Research Paper No. 1. San Diego.

Smith E.A.
1988 Risk and Uncertainty in the "Original Affluent Society:" Evolutionary Ecology of Resource Sharing and Land Tenure. In *Hunters and Gatherers: History, Evolution, and Social Change*, edited by Ingold, Riches, and Woodburn, pp. 222-252. Berg, Oxford.

Soto, A. (O.F.M.)
1961 Mission San Luis Rey, California: Excavations in the Sunken Gardens. *Kiva* 26(4):34-36.

Sparkman, P.S.
1905 Sketch of the Grammar of the Luiseno Language of California. *American Anthropologist* 7(4):656-662.

1908a The Culture of the Luiseno Indians. *University of California Publications in American Archaeology and Ethnology* 8(4):187-234. Berkeley.

1908b Notes on California Folk-Lore: A Luiseno Tale. *Journal of American Folk-Lore* 21(80):35-36.

Steward, J.
1938 Basin Plateau Aboriginal Sociopolitical Groups. *Bureau of American Ethnology Bulletin* 120. Smithsonian Institution, Washington D.C.

1955 *Theory of Culture Change*. University of Illinois Press, Champaign.

Stuiver, M., G.W. Pearson and T. Branziunas
1986 Radiocarbon Age Calibration of Marine Samples Back to 9000 Calendar Years B.P. *Radiocarbon* 28(2B):980-1021.

Strong, W.D.
1929 Aboriginal Society in Southern California. *University of California Publications in American Archaeology and Ethnology* 26.

Taylor, R.E.
1987 *Radiocarbon Dating*. Academic Press, New York.

Taylor, R.E. and R. Berger
1967 Radiocarbon Content of Marine Shells from the Pacific Coasts of Central and South America. *Science* 158:1180-1182.

Taylor, R.E., P.J. Ennis, L.A. Payen, C.A. Prior and P.J. Slota, Jr.
1986 Encino Village (CA-LAn-43) Site Radiocarbon Determinations: Geophysical/Geochemical Considerations. *Pacific Coast Archaeological Society Quarterly* 22(3):35-48.

Thomas, D.H.
1972 A Computer Simulation of Great Basin Shoshonean Subsistence and Settlement Patterns. In *Models in Archaeology*, edited by D.L. Clarke. Methuen, London.

1973 An Empirical Test for Steward's Model of Great Basin Settlement Patterns. *American Antiquity* 38:155-176.

Tilley, C.
1985 Archaeology as Sociopolitical Action in the Present. In *Critical Traditions in Contemporary Archaeology*, edited by V. Pinsky and A. Wylie. Cambridge University Press, Cambridge.

Tordoff, Judith
1991 Personal Communication. Caltrans Archaeologist, Sacramento.

Towner, Ronald H.
1991 Chapter 9 Admiralty Site Lithic Analysis. In *Life in The Ballona: Cultural Resources Investigations at The Admiralty Site (CA-LAN-47) and The Channel Gateway Site (CA-LAN-1596-H)* by Jeffrey H. Altschul and Jeffrey A. Homburg, pp. 206-256. Statistical Research Technical Series No. 29. Tucson.

Treganza, A.E. and A. Bierman
1958 The Topanga Culture: Final Report on Excavation, 1948. *University of California Anthropological Records* 20:45-86. Berkeley and Los Angeles.

Troxell, H.C.
1942 *Floods of March 1938 in Southern California*. U.S. Geological Survey, Washington, D.C.

True, D.L.
1954 Pictographs of the San Luis Rey Basin, California. *American Antiquity* 20(3):255-264.

1957 Fired Clay Figurines from San Diego County, California. *American Antiquity* 22:291-296.

1958 An Early Complex in San Diego County, California. *American Antiquity* 23(3):255-264

1966 *Archaeological Differentiation of Shoshonean and Yuman Speaking Groups in Southern California.* Unpublished Ph.D. dissertation. Department of Anthropology, University of California, Los Angeles.

1970 *Investigation of a Late Prehistoric Complex in Cuyamaca Rancho State Park, San Diego County, California*. University of California, Los Angeles, Archaeological Survey Monographs No. 1. Los Angeles.

1977 *Archaeological Investigations in San Diego County, California. Preliminary Report on Sites SDi-4558, 4562.* Report prepared for the California Department of Transportation, Sacramento.

1980 The Pauma Complex in Northern San Diego County: 1978. *The Journal of New World Archaeology* 3(4):1-39.

1983 Casual Artifacts in Northern San Diego County, California: The Hammergrinder. *Journal of California and Great Basin Anthropology* 5(1 and 2): 208-223.

1985 Ovate Pestles and an Isoloated Processing Station in Interior Southern California. *Journal of California and Great Basin Anthropology* 7(2): 254-260.

True, D.L. and Eleanor Beemer
1982 Two Milling Stone Inventories from Northern San Diego County, California. *Journal of California and Great Basing Anthropology* 4(2):233-261.

True, D.L. and R. Pankey
1985 Radiocarbon Dates for the Pauma Complex Component at the Pankey Site, Northern San Diego County, California. *Journal of California and Great Basin Anthropology* 7(2):240-244.

True D.L. and G. Waugh

1981 Archaeological Investigations in Northern San Diego County, California: Frey Creek. *Journal of California and Great Basin Anthropology* 3(1):84-115

1982 Proposed Settlement Shifts during San Luis Rey Times: Northern San Diego County, California. *Journal of California and Great Basin Anthropology* 4(2):34-54.

True, D.L., C.W. Meighan and H. Crew

1974 *Archaeological Investigations at Molpa, San Diego County, California.* University of California Publications in Anthropology No. 11.

True, D.L., Rosemary Pankey and C.N. Warren

1991 *Tom-Kav: A Late Village Site in Northern San Diego County, California, and Its Place in the San Luis Rey Complex.* University of California Publications Anthropological Records, Volume 30, University of California Press.

Udvardy, Miklos D.F.

1984 *The Audubon Society Field Guide to North American Birds*. Alfred A. Knoph, Inc., New York.

United States Department of the Interior

1979 *An Ecological Characterization Study of the Chenier Plain Coastal Ecosystem of Louisiana and Texas*, Vol.1, Narrative Report. Office of Research and Development, U.S. Environmental Protection Agency, and Fish and Wildlife Service.

Van Devender, T.R. and W.G. Spaulding

1979 Development of Vegetation and Climate in the Southwestern United States. *Science* 204:701-710.

Van Horn, D.

1983 Cultural Resource Report Form, County of San Diego: CA-SDI-5992. Ms. on file, County of San Diego, Department of Parks and Recreation, San Diego.

Vizcaino, J.

1959 *The Sea Diary of Fr. Juan Vizcaino to Alta California 1769.* A. Woodward (translator). Glenn Dawson, Los Angeles.

Wallace, W.J.

1955 A Suggested Chronology for Southern California Coastal Archaeology. *Southwestern Journal of Anthropology* 11:214-230.

1962 Prehistoric Cultural Development in the Southern California Deserts. *American Antiquity* 28(2):172-180.

1978 Post-Pleistocene Archaeology. In *California*, edited by R.F. Heizer, pp. 25-36. Handbook of North American Indians, vol. 8, William G. Sturtevant, general editor. Smithsonian Institution, Washington, D.C.

Warren, C.N.

1964 *Cultural Change and Continuity on the San Diego Coast.* Unpublished Ph.D. dissertation, Department of Anthropology, University of California, Los Angeles.

1966 *The San Dieguito Type Site: M.J. Rogers' 1938 Excavation on the San Dieguito River.* San Diego Museum of Man Papers No. 5. San Diego.

Warren, C.N.
1967 The San Dieguito Complex: A Review and Hypothesis. *American Antiquity* 32(2):233-236.

1968 Cultural Tradition and Ecological Adaptation on the Southern California Coast. In, *Archaic Prehistory in the Western United States*, edited by C. Irwin-Williams, pp. 1-14. Eastern New Mexico University Contributions in Anthropology No. 1. Portales

1987 San Dieguito and La Jolla: Some Comments. In *San Dieguito - La Jolla: Chronology and Controversy*, edited by D.R. Gallegos and S.M. Hector, pp. 75-85. San Diego County Archaeological Society Research Paper No. 1.

Warren, C.N., and M.G. Pavesic
1963 Shell Midden Analysis of Site SDi-603 and Ecological Implications for Cultural Development of Batiquitos Lagoon, San Diego County. *University of California, Los Angeles, Archaeological Survey Annual Report*, 1962-1963: 407-438. Los Angeles.

Warren, C.N. and D.L. True
1961 The San Dieguito Complex and Its Place in California Prehistory. In *University of California, Los Angeles, Archaeological Survey Annual Report, 1960-1961:* 246-338. Los Angeles.

Warren, C.N., D.L. True and A.A. Eudey
1961 *Early Gathering Complexes of Western San Diego County: Results and Interpretations of an Archaeological Survey*. University of Illinois Press, Champaign.

Warter, Janet K.
1976 Late Pleistocene Plant Communities - Evidence from the Rancho La Brea Tar Pits. In *Plant Communities of Southern California,* edited by June Latting, pp. 32-39. California Native Plant Society Special Publication No. 2. Los Angeles.

Watson, P.J., S.A. LeBlanc and C.L. Redman
1971 *Explanation in Archeology: An Explicitly Scientific Approach*. Columbia University Press, New York

1984 *Archeological Explanation: The Scientific Method in Archeology.* Columbia University Press, New York.

Weide, M.L.
1968 *Cultural Ecology of Lakeside Adaptation in the Western Great Basin.* Unpublished Ph.D. dissertation. Department of Anthropology, University of California, Los Angeles.

Wells, H. and N. Farrell
1990a *Research Design for Test Excavations at SDi-6013, 6014, 6015, 5130, and 5133, San Luis Rey River, Oceanside, California.* Prepared for Los Angeles District, Corps of Engineers, Los Angeles.

1990b *Preliminary Investigation of Sites CA-SDi-5130, -5133, -6013, -6014, and -6015, San Luis Rey River, Oceanside, San Diego County, California.* Prepared for Los Angeles District, Corps of Engineers, Los Angeles.

White, R.C.
1963 Luiseno Social Organization. *University of California Publications in American Archaeology and Ethnology* 48(2):91-194.

Willey, G.R. and J.A. Sabloff
1980 *A History of American Archaeology*. Second edition. W.H. Freeman and Company, San Francisco.

Wolff, W.J.
1983 Estuarine Benthos. In *Ecosystems of the World, 26: Estuaries and Enclosed Seas,* edited by B.H. Ketchum, pp. 151-182. Elsevier Scientific Publishing Company, New York.

Woodward-Clyde Consultants
1990 Geologic Reconnaissance Study, Whelan Dairy, Oceanside, California. In *Draft Environmental Impact Report for the Whelan Property Acquisition*. Affinis, 1991.

Yellen, J.
1977 *Anthropological Approaches to the Present*. Academic Press, New York.

Yesner, D.R.
1980 Maritime Hunter-gatherers: Ecology and Prehistory. *Current Anthropology* 21:727-735.